THE APPROXIMATE MINIMIZATION OF FUNCTIONALS

Prentice-Hall Series in Automatic Computation

George Forsythe, *editor*

ARBIB *Theories of Abstract Automata*
BATES AND DOUGLAS *Programming Language/One, 2nd edition*
BAUMANN, FELICIANO, BAUER, AND SAMELSON *Introduction to ALGOL*
BLUMENTHAL *Management Information Systems*
BORROW AND SCHWARTZ, editors *Computers and the Policy-Making Community: Applications to International Relations*
BOWLES, editor *Computers in Humanistic Research*
CRESS, DIRKSEN, AND GRAHAM *FORTRAN IV with WATFOR and WATFIV*
DANIEL *The Approximate Minimization of Functionals*
DESMONDE *Computers and Their Uses, 2nd edition*
EVANS, WALLACE, AND SUTHERLAND *Simulation Using Digital Computers*
FIKE *Computer Evaluation of Mathematical Functions*
FORSYTHE AND MOLER *Computer Solution of Linear Algebraic Systems*
GAUTHIER AND PONTO *Designing Systems Programs*
GORDON *System Simulation*
GREENSPAN *Lectures on the Numerical Solution of Linear, Singular and Nonlinear Differential Equations*
HARTMANIS AND STEARNS *Algebraic Structure Theory of Sequential Machines*
HULL *Introduction to Computing*
JOHNSON *System Structure in Data, Programs, and Computers*
LORIN *Parallelism in Hardware and Software: Real and Apparent Concurrency*
MARTIN *Design of Real-Time Computer Systems*
MARTIN *Future Developments in Telecommunications*
MARTIN *Programming Real-Time Computer Systems*
MARTIN *Telecommunications and the Computer*
MARTIN *Teleprocessing Network Organization*
MARTIN AND NORMAN *The Computerized Society*
MATHISON AND WALKER *Computers and Telecommunications: Issue in Public Policy*
MCKEEMAN, HORNING, AND WORTMAN *A Compiler Generator*
MINSKY *Computation: Finite and Infinite Machines*
MOORE *Interval Analysis*
PYLYSHYN, editor *Perspectives on the Computer Revolution*
SALTON *The SMART Retrieval System: Experiments in Automatic Data Processing*
SAMMET *Programming Languages: History and Fundamentals*
STERLING AND POLLACK *Introduction to Statistical Data Processing*
TAVISS, editor *The Computer Impact*
TRAUB *Iterative Methods for the Solution of Equations*
VARGA *Matrix Iterative Analysis*
VAZSONYI *Problem Solving by Digital Computers with PL/1 Programming*
WILKINSON *Rounding Errors in Algebraic Processes*

Prentice-Hall International, Inc., *London*
Prentice-Hall of Australia, Pty. Ltd., *Sydney*
Prentice-Hall of Canada, Ltd., *Toronto*
Prentice-Hall of India Private Limited, *New Delhi*
Prentice-Hall of Japan, Inc., *Tokyo*

THE APPROXIMATE MINIMIZATION OF FUNCTIONALS

JAMES W. DANIEL

Computer Sciences Department
University of Wisconsin

PRENTICE-HALL, INC.

ENGLEWOOD CLIFFS, N. J.

Current printing (first digit):
1 2 3 4 5 6 7 8 9 10

13-043877-4

Library Of Congress Catalog Card Number: 70-133091

Printed in the United States of America

TO R. J. BRYANT, P. T. MIELKE,
G. E. FORSYTHE, G. H. GOLUB, AND M. M. SCHIFFER,
WITH MANY THANKS

PREFACE

This material was prepared for use as a text in support of my lectures at the 1969 session on "Minimization problems" of l'Ecole d'Eté Analyse Numérique held at Le Breau-sans-Nappe, France, and sponsored by the Commissariat á l'Energie Atomique, Gaz de France, and l'Electricité de France; I gratefully acknowledge the financial support of those organizations and their cooperation in reproducing this text. Of course those ideas in this text which are primarily the author's did not come to light only during the period of writing; therefore I must also acknowledge several other sources of support in recent years, particularly the National Science Foundation, the Office of Naval Research, the Mathematics Research Center at the University of Wisconsin, and the Computer Sciences Department at the University of Wisconsin, Madison. It is a great pleasure for me to acknowledge the invaluable cooperation of my three assistants at l'Ecole d'Eté, Mssrs. Chavent, Yvon, and Tremolières, and the students there, especially Mr. Lascaux, whose questions, comments, and suggestions were of great help to me.

This book represents an attempt at a general presentation of problems involved in the approximate minimization of functionals. As such it treats in the first chapter the general area of variational problems and the question of the existence of solutions, in the next two chapters the question of the convergence of approximate solutions of discretized variational problems in general and in specific important cases, in the next two chapters the theory of gradient methods of minimization in general spaces, and in the final four chapters practical computational methods of minimization in $\mathbb{R}^l$. Although a significant amount of material is presented concerning constrained minimization problems, the primary emphasis of this text is on unconstrained problems, particularly in the last four chapters; this is not a text on mathematical programming. In the last four chapters on "practical" methods, detailed computational algorithms or computer programs are not presented. The rather "theoretical" presentation of this "practical" material reflects my feeling that analysis and understanding of an algorithm, even under specialized hypotheses, is all-important in the creation of new methods; whenever possible, however, the references include sources of computer programs or detailed computational descriptions of the implementations of algorithms.

The references presented here give a reasonably thorough coverage of the most important English language literature on minimization methods, although no claim is made of completeness. Although many of the important results of Russian papers can also be found in English, the largest gaps in the references are with respect to the very extensive literature in Russian. I hope to correct this shortcoming at some future time.

JAMES W. DANIEL

Madison, Wisconsin

CONTENTS

4 GENERAL BANACH-SPACE METHODS OF GRADIENT TYPE 70

5 CONJUGATE-GRADIENT AND SIMILAR METHODS IN HILBERT SPACE 114

6 GRADIENT METHODS IN $\mathbb{R}^l$ 142

7 VARIABLE-METRIC GRADIENT METHODS IN $\mathbb{R}^l$ 159

1 VARIATIONAL PROBLEMS IN AN ABSTRACT SETTING

1.1. INTRODUCTION

Many problems of pure and applied mathematics either arise or can be formulated as *variational* problems, that is, as problems of locating a minimizing point for some (nonlinear) real-valued functional over a certain set. Such a setting is often beneficial from the analytic viewpoint of determining the existence and uniqueness of such points; however, we shall primarily emphasize here the computational aspects of this approach, that is, how we can compute a solution to the problem by actually minimizing the appropriate functional. In some cases, in which the problem was originally formulated as a variational problem (e.g., calculus-of-variations problems), this may well be one of the best computational approaches, rather than considering the problem in some equivalent form (e.g., the Euler–Lagrange differential equation). In other cases, in which the problem is artificially converted into a variational one, this approach provides new computational methods for consideration.

General references: Kantorovich–Krylov (1958), Courant–Hilbert (1953), Kantorovich (1948).

1.2. TYPICAL PROBLEMS

The variety of minimization problems is indeed immense; we consider some typical problems. A large class is that of *optimal-control* problems, which arise so often in the highly technological industries. Here one seeks to minimize a cost functional

$$f(x, u) = \int_0^{t_F} c[t, x(t), u(t)]\, dt$$

over the collection of points (functions) x, u satisfying constraints such as

$$\dot{x} \equiv \frac{dx}{dt} = s[t, x(t), u(t)]$$
$$x(0) \in X_I$$
$$x(t_F) \in X_F$$
$$x(t) \in X(t)$$
$$u(t) \in U(t)$$

where X_I, X_F, $X(t)$, and $U(t)$ are certain specified sets and set functions. Such problems arise, for example, in determining the minimal time or minimal expenditure for striking a given missile target.

As a special case with $s(t, x, u) \equiv u$, we have a basic problem of the *calculus of variations*, in which we seek to minimize

$$f(x) = \int_0^1 c[t, x(t), \dot{x}(t)]\, dt$$

subject to some boundary conditions on $x(t)$, such as

$$x(0) = x(1) = 0$$

Many problems of applied mathematics are of this latter form if we allow the variable t to represent a vector of dimension higher than one and $\dot{x}$ to represent the vector of first partial derivatives of x with respect to those variables. The classical problem of finding the height $x(t_1, t_2)$ of a surface of minimal area stretched across a closed space curve $C = c(t_1, t_2)$ for $(t_1, t_2) \in G$, a simple closed curve in the plane, for example, is most naturally described mathematically as the problem of finding $x(t_1, t_2)$ to minimize

$$\int_{\hat{G}} \left[1 + \left(\frac{\partial x}{\partial t_1}\right)^2 + \left(\frac{\partial x}{\partial t_2}\right)^2\right]^{1/2} dt_1\, dt_2 \qquad (\hat{G} = \text{region enclosed by } G)$$

over the set of $x(t_1, t_2)$ satisfying

$$x(t_1, t_2) = c(t_1, t_2) \quad \text{for} \quad (t_1, t_2) \in G$$

A discrete analogue of the continuous optimal-control problem described above is the *mathematical-programming* problem in which one seeks to minimize some function of a finite number of real variables subject to finitely many constraints; these problems arise, for example, very commonly in the petroleum industry. As we shall later see, it is also often solved as a discretized approximation to a continuous optimal-control problem. Many problems in data approximation ultimately are of this form also and provide perhaps one of the most common forms of such problems. Where one has some data y_i depending on a parameter t measured at certain points t_i, $i = 1, 2, \ldots, N$, it is desired to approximate the formula generating the data by some expression

$$y(t) \simeq g(t; \alpha)$$

where the choice of the parameter α determines the expression. We can do this by picking α so as to minimize some norm of the vector in E^N with components

$$y_i - g(t_i; \alpha)$$

Essentially, this technique has been applied lately in a somewhat novel fashion for the solution of such problems as differential or integral equations. For example, one seeks to solve the differential equation

$$Du = f \quad \text{in} \quad C, \qquad u = 0 \quad \text{on} \quad \partial C = \text{boundary of } C$$

where C is some domain. If $\varphi_1, \ldots, \varphi_N$ are some functions satisfying the boundary condition on ∂C, we try to choose numbers $\alpha_1, \ldots, \alpha_N$ to minimize some norm of the vector in E^M with components given by

$$[Du](t_i) - f(t_i), \qquad i = 1, \ldots, M$$

where $\{t_i\}$ is some grid of points over the domain C; $\alpha_1\varphi_1 + \cdots + \alpha_N\varphi_N$ is then taken as an approximate solution to the differential equation.

This represents only a few of the types of variational problems commonly of practical interest at present; rather than proceed to list more types, we shall develop tools to allow us to consider such problems in a more general setting.

General references: Hestenes (1966), Pontryagin et al. (1962), Balakrishnan–Neustadt (1964), Courant–Hilbert (1953), Morrey (1966), Akhiezer (1962), Abadie (1967), Fiacco–McCormick (1968), Hadley (1964), Mangasarian (1969), Zangwill (1969), Lorentz (1966), Rosen–Meyer (1967), Rabinowitz (1968), Mikhlin–Smolitskiy (1967).

1.3. BASIC FUNCTIONAL ANALYSIS

We shall consider variational problems which can be stated as problems in a general function-space setting; for simplicity, we restrict ourselves to problems in Banach spaces. We recall that a real or complex *Banach space E* (hereafter assumed to be real unless otherwise stated) is a vector space over the real or complex numbers which is complete in the topology generated by the norm $||\cdot||$ defined on E. A *Hilbert space* is a Banach space in which the norm is given via an inner product $\langle\cdot, \cdot\rangle$, that is, $||x|| = \langle x, x\rangle^{1/2}$.

The Banach space E^* of bounded (real-valued) linear functionals on E is called the *dual* or *adjoint* space; it generates another topology for E, the *weak* topology, for which a basis at the origin 0 is given by

$$\{\mathcal{O};\mathcal{O} = \{x; |f_1(x)| < a_1, \ldots, |f_n(x)| < a_n\}, f_1, \ldots, f_n \in E^*, a_1, \ldots, a_n > 0\}$$

Convergence to a point x of a sequence $\{x_n\}$ in this topology, that is, weak convergence, is denoted by

$$x_n \rightharpoonup x$$

Convergence in the norm topology, for which $\lim_{n\to\infty} ||x_n - x|| = 0$, is denoted by

$$x_n \longrightarrow x$$

Recall that $x_n \rightharpoonup x$ if and only if $\lim_{n\to\infty} |f(x_n) - f(x)| = 0$ for each fixed $f \in E^*$.

We can consider E as a subset of E^{**} since, for each fixed x in E, we can define a bounded linear functional over E^* via

$$x(f) \equiv f(x) \quad \text{for all } f \text{ in } E^*$$

If $E = E^{**}$ under this identification, E is said to be *reflexive*. Recall that a *compact* (*countably compact*) set is one such that every open cover (countable open cover) has a finite subcover, and a sequentially compact set C is one such that every sequence in C has a subsequence converging to a point of C. Every bounded, weakly closed subset of reflexive space is weakly sequentially compact; the important theorem of Alaoglu moreover says that every such set is weakly compact.

Since every continuous linear functional f on a Hilbert space E is given by an inner product—that is, there exists a unique y in E with $f(x) = \langle x, y\rangle$ for all x—we can write $E = E^*$, which in particular implies that E is reflexive.

Given a finite set of points $\{x_1, \ldots, x_n\}$ in E, a *convex combination* of the points is a point x such that

$$x = \sum_{i=1}^{n} a_i x_i, \qquad \sum_{i=1}^{n} a_i = 1, \qquad a_i \geq 0, \qquad i = 1, 2, \ldots, n$$

A set C in E is said to be *convex* if every convex combination of every finite subset of C is itself an element of C. The following important result is known: If $x_n \rightharpoonup x$, then for each $n = 1, 2, \ldots$, there is a convex combination y_n of the points $x_1, \ldots, x_n$ such that $y_n \longrightarrow x$. This is implied by the fact that a convex set which is closed in the norm topology must be closed in the weak topology; a weakly closed set is always norm-closed of course since a norm convergent sequence is weakly convergent.

EXERCISE. Prove that the fact that a convex set is norm-closed if and only if it is weakly closed implies the fact that for every weakly convergent sequence $\{x_n\}$, there is a sequence $\{y_n\}$ of convex combinations of $\{x_n\}$ that norm-converges to the same limit.

In the sequel we shall be considering (real-valued) nonlinear functionals f on a space E and various continuity properties of those functionals. Recall that a functional f is *lower semicontinuous* in a topology if and only if $\{x; f(x) > a\}$ is open for each real a. From this we deduce that if f is *weakly lower semicontinuous*—that is lower semicontinuous in the weak topology—then from $x_n \rightharpoonup x$ it follows that

$$f(x) \leq \liminf_{n\to\infty} f(x_n)$$

A similar statement holds for norm lower semicontinuity, of course, under the hypothesis that $x_n \longrightarrow x$. Since most of our arguments depend only on the above sequential property, we define a functional to be *sequentially lower semicontinuous* (for some topology) if and only if the convergence of $\{x_n\}$ to x implies $f(x) \leq \liminf_{n\to\infty} f(x_n)$. Sequential continuity properties are generally weaker than the usual continuity properties; the two kinds of properties are equivalent in the norm topology since the norm topology has a countable basis at each point.

EXERCISE. Prove that a functional on a normed space is norm sequentially lower semicontinuous if and only if it is norm lower semicontinuous.

We shall often consider linear mappings from one Banach space E_1 into another E_2; if A is such a mapping, then it is continuous (between the norm topologies) if and only if it is *bounded*, that is,

$$||A|| \equiv \sup_{||x||_{E_1}=1} ||Ax||_{E_2} = \sup_{x\in E_1} \frac{||Ax||_{E_2}}{||x||_{E_1}} < \infty$$

The Banach space of all such bounded linear operators is denoted by $L(E_1, E_2)$; thus $E^* = L(E, \mathbb{R})$. If $A \in L(E_1, E_2)$ and $f_2 \in E_2^*$, we can define $f_1 \in E_1^*$ by

$$f_1(x) = f_2(Ax)$$

The mapping of f_2 into f_1 is linear and is denoted by A^*, the *adjoint* of A. If E is a Hilbert space, and if $A \in L(E, E)$ and $A^* = A$, then A is called *self-adjoint*; in this case we have $\langle Ax, y\rangle = \langle x, A^*y\rangle = \langle x, Ay\rangle$ for all $x, y \in E$.

If E is a real (complex) Banach space and $A \in L(E, E)$, then $\sigma(A)$, the *spectrum* of A, is defined via

$$\sigma(A) = \{\lambda \in \mathbb{R}\,(\mathbb{C});\; A_\lambda \equiv \lambda I - A \quad \text{is not one to one,} \\ \text{or} \quad \overline{R(A_\lambda)} \neq E, \quad \text{or} \quad A_\lambda^{-1} \notin L[\overline{R(A_\lambda)}, E]\}$$

where $R(A_\lambda)$ denotes the range of A_λ and $\bar{S}$ denotes the norm closure of any set S. If E is a Hilbert space and A is self-adjoint, then all points in $\sigma(A)$ are real and lie in the interval $[m, M]$ where

$$m = \inf_E \frac{\langle x, Ax\rangle}{\langle x, x\rangle} \quad \text{and} \quad M = \sup_E \frac{\langle x, Ax\rangle}{\langle x, x\rangle}$$

Then we can write

$$mI \leq A \leq MI$$

meaning

$$m\langle x, x\rangle \leq \langle x, Ax\rangle \leq M\langle x, x\rangle$$

It follows that $||A|| = \max(|m|, |M|)$. If $m > 0\ (\geq 0)$, A is called *positive-definite* (*semidefinite*) or *coercive* (*semicoercive*).

Examples of simple spaces

1. $E = \mathbb{R}^n$, n-dimensional real Euclidean space with any norm whatsoever since all norms on $\mathbb{R}^n$ are topologically equivalent. Some standard norms are, for $x = (x_1, \ldots, x_n)$:

$$||x||_{n,\infty} = \max_{1 \leq i \leq n} |x_i| \quad \text{and} \quad ||x||_{n,p} = \left[\sum_{i=1}^{n} |x_i|^p\right]^{1/p}$$

in the second equation particularly for $p = 1$ or $p = 2$, in which latter case E is a Hilbert space.

2. $E = C[a, b]$ is the real Banach space of real-valued continuous functions defined on $[a, b] \subset \mathbb{R}$ with the norm $||f||_\infty = \max_{[a,b]} |f(t)|$; norm convergence here is equivalent to the usual uniform convergence. E is not reflexive.

3. $E = L_p(a, b)$, the Banach space of real-valued pth-power Lebesgue integrable (if $p = \infty$, then essentially bounded instead) functions defined on $(a, b) \subset \mathbb{R}$, with norm

$$||f||_p \equiv \left[\int_a^b |f(t)|^p\, dt\right]^{1/p}$$

or

$$||f||_\infty \equiv \operatorname*{ess\,sup}_{(a,b)} |f(t)|$$

For $1 < p < \infty$, $L_p(a, b)$ is reflexive; $L_2(a, b)$ is a Hilbert space.

4. $E = W_p^m(a, b)$, the Banach space of all real-valued functions defined on $[a, b]$, having absolutely continuous derivatives of orders less than or equal to $(m - 1)$ on $[a, b]$, and having the mth derivative (and of course all

lower-order derivatives) in $L_p(a, b)$. Many equivalent norms are possible for this *Sobolev space*, one of which is

$$||f||_p^m = \left\{\sum_{i=0}^{m-1} |f^{(i)}(a)|^p + \int_a^b |f^m(t)|^p\, dt\right\}^{1/p}$$

If $p = 2$, W_2^m is a Hilbert space with the obvious inner product. (Such spaces can be defined in higher dimensions, although the concept of the derivatives of absolutely continuous functions must be replaced by that of generalized or distributional derivatives, and the norm should be the sum of the L_p norms of all the derivatives of order i for $0 \leq i \leq m$.) For clarity we consider the Hilbert space $W_2^1(0, 1)$. Here we have

$$||f||_2^1 = [\langle f, f\rangle_2^1]^{1/2} \equiv \left\{|f(0)|^2 + \int_0^1 |f'(t)|^2\, dt\right\}^{1/2}$$

Thus, if $f_n \longrightarrow f$ in W_2^1, then f_n' converges to f' in $L_2(0, 1)$ and f_n converges to f uniformly, that is, in $C[0, 1]$. Even weak convergence in W_2^1 implies a strong type of convergence. For, if $f_n \rightharpoonup f$ in W_2^1, then $f_n(0)$ converges to $f(0)$ and f_n' converges to f' weakly in $L_2(0, 1)$. From this it follows that $\{f_n\}_1^\infty$ is equicontinuous and uniformly bounded, which implies that f_n converges to f uniformly on $[0, 1]$, that is, it converges in $C[0, 1]$.

EXERCISE. Provide the details proving that weak convergence in W_2^1 implies norm convergence in $C[0, 1]$.

Since many algorithms yield sequences only weakly convergent to some desired solution, it is important to realize, as this example indicates, that weak convergence can sometimes be quite "strong."

General references: Dunford–Schwartz (1962), Kantorovich (1948), Kantorovich–Akilov (1964), Morrey (1966), Taylor (1961).

1.4. GENERAL FUNCTIONAL ANALYSIS FOR MINIMIZATION PROBLEMS

Perhaps the fundamental result concerning the existence of a solution for a variational problem is the well-known general fact that a sequentially lower semicontinuous real-valued function defined on a countably compact set must achieve its infimum there; the important special case for our use is the following.

THEOREM 1.4.1. Let f be a weakly sequentially lower semicontinuous functional defined on a weakly sequentially compact set C; then there exists x_0 in C such that $f(x_0) = \inf_{x \in C} f(x) = \min_{x \in C} f(x)$.

Proof: Let $m = \inf_{x \in C} f(x)$; then there exists a sequence of points $x_n \in C$, with $\lim_{n \to \infty} f(x_n) = m$. Since C is weakly sequentially compact, there is an $x_0 \in C$ and a subsequence $x_{n'}$ such that $\lim_{n' \to \infty} f(x_{n'}) = m$ and $x_{n'} \rightharpoonup x_0$. By the semicontinuity,

$$f(x_0) \leq \liminf_{n' \to \infty} f(x_{n'}) = m \leq f(x_0)$$

Thus $f(x_0) = m$. Q.E.D.

COROLLARY 1.4.1. A weakly sequentially lower semicontinuous functional achieves its infimum on every convex, bounded, norm-closed subset of a reflexive space.

Proof: From earlier statements we know that a convex, norm-closed set is weakly closed and that a weakly closed, bounded set is weakly sequentially compact in reflexive space; the corollary then follows immediately from the theorem. Q.E.D.

EXERCISE. Prove that a lower semicontinuous functional achieves its infimum on each countably compact set.

Theorem 1.4.1, of course, directly applies only to *constrained* minimization problems, that is, where x is considered only in some weakly sequentially compact, and hence bounded, set C; our main concern here is with unconstrained problems, but for proving existence of a solution it is often possible and wise to reduce them to constrained problems. (An interesting sidelight is that for *computational* purposes one often tries to reduce the constrained problem to an unconstrained one.) Here one tries to find a bound—say B—on the norm of the solution if it exists, and then consider the problem over $S_B = \{x; ||x|| \leq B\}$. This is possible if, for example, the functional satisfies a T-property in a reflexive space.

DEFINITION 1.4.1. A functional f is said to satisfy a *T-property* (at x_0 with $T = T_0$) if there exists an x_0 and $T_0 > 0$ such that $||x - x_0|| \geq T_0$ implies $f(x) > f(x_0)$.

THEOREM 1.4.2. A weakly sequentially lower semicontinuous functional f satisfying a T-property in a reflexive space E achieves its infimum over E.

Proof: Let $S(x_0; T_0)$ be the norm-closed sphere of radius T_0 about x_0; then $S(x_0; T_0)$ is weakly sequentially compact. Since $f(x_0) < f(x)$ if $x \notin S(x_0; T_0)$, we have

$$\inf_{x \in S(x_0; T_0)} f(x) \leq f(x_0) \leq \inf_{x \notin S(x_0; T_0)} f(x)$$

Thus

$$\inf_{x \in S(x_0; T_0)} f(x) = \inf_{x \in E} f(x)$$

By Theorem 1.4.1 there exists $x' \in S(x_0; T_0)$ achieving the infimum over $S(x_0; T_0)$ and hence over E. Q.E.D.

Satisfaction of a T-property is usually guaranteed by a growth property of the functional; for example, if $\lim_{||x|| \to \infty} f(x) = +\infty$, then f certainly satisfies a T-property. In the next section we shall see that various convexity assumptions can lead to such growth properties as well as the needed semicontinuity property.

One important application of and reason for studying variational problems lies in the study of the solution of operator equations; this arises from the generalization of the calculus result that if a differentiable real-valued function of several real values achieves a local extremum at an interior point x_0 of a set, then the gradient of the functional vanishes at x_0. For this and other reasons it is necessary to consider the concept of the derivative of a general operator.

Let F map a Banach space E_1 into a Banach space E_2. Although a weaker concept of the derivative—namely, the Gateaux derivative or differential—is often of interest, for our purposes we need only consider the stronger concept.

DEFINITION 1.4.2. If at a point x in E_1, there exists a bounded linear operator denoted F'_x mapping h in E_1 into $F'_x h$ in E_2—that is, $F'_x \in L(E_1, E_2)$—and satisfying

$$\lim_{||h|| \to 0} \frac{||F(x + h) - F(x) - F'_x h||}{||h||} = 0$$

then F'_x is called the (*Frechet*) *derivative* of F at x.

It is simple to prove that the properties of the Frechet derivative are analogous to those statements in the standard calculus.

PROPOSITION 1.4.1. If the Frechet derivative f'_x of a real-valued functional f exists at each point of a convex set C, then for $x \in C$ and $x + h \in C$, the Lagrange formula

$$f(x + h) - f(x) = f'_{x+th} h$$

is valid for some $t \in (0, 1)$. If the operator F from E_1 into E_2 has a Frechet derivative F'_x at each point x in a convex set C, then for $x \in C$ and $x + h \in C$, the Lipschitz condition

$$||F(x+h) - F(x)|| \leq ||F'_{x+th}|| \, ||h||$$

is valid for some $t \in (0, 1)$. If F'_x exists at some point x, then F is norm continuous at x. If F maps E_1 into E_2 and G maps E_2 into E_3 with F and G Frechet-differentiable, then GF is Frechet-differentiable and $(GF)'_x = G'_{F(x)}F'_x$. The usual rules $(F_1 + F_2)' = F'_1 + F'_2$ and $(cF)' = cF'$ are also valid under the obvious assumptions.

EXERCISE. Prove the statements in Proposition 1.4.1.

If f is a functional over E, then f'_x is a bounded linear operator into $\mathbb{R}$, that is, $f'_x \in E^*$; in such a case f'_x is called the *gradient* of f at x and is denoted

$$f'_x \equiv \nabla f(x) = (\nabla f)(x)$$

Thus we can write

$$f(x+h) = f(x) + \langle h, \nabla f(x) \rangle + o(h) \quad \text{as} \quad ||h|| \longrightarrow 0$$

where

$$\langle h, g \rangle \equiv g(h) \quad \text{for} \quad g \in E^*, \qquad h \in E$$

Since $\nabla f = f'$ is an operator from E into E^* we can define its derivative (the second derivative of f) and so on through derivatives of arbitrary order; f''_x, for example, then maps elements h of E into $f''_x h$ in E^*.

It is easy to find conditions in terms of derivatives which will imply the growth and continuity properties important for variational problems. Although we shall return to this later with a stronger result (Theorem 1.5.1), we give a typical theorem below.

THEOREM 1.4.3. Suppose the functional f has first and second derivatives at all points of a convex set C, and that for x in C, f''_x satisfies $\langle h, f''_x h \rangle \geq 0$ for all h in E. Then f is weakly sequentially lower semicontinuous in C.

Proof: Let $x \in C$, $x_n \rightharpoonup x$, $x_n \in C$. Then by the Lagrange property,

$$f(x_n) = f(x) + \langle x_n - x, \nabla f(x + th_n) \rangle \quad \text{for some } t \in (0, 1)$$

with $h_n = x_n - x$. Then

$$f(x_n) = f(x) + \langle x_n - x, \nabla f(x) \rangle + \langle h_n, \nabla f(x + th_n) - \nabla f(x) \rangle$$

By the Lagrange property for the function

$$g(\tau) \equiv \langle h_n, \nabla f(x + \tau th_n) - \nabla f(x) \rangle, \qquad \tau \in [0, 1]$$

it is easy to see that

$$f(x_n) = f(x) + \langle x_n - x, \nabla f(x) \rangle + \langle h_n, f''_{x+\tau t h_n} t h_n \rangle$$

for some point (t, τ) in $(0, 1) \times (0, 1)$. Since $x_n \rightharpoonup x$, $\langle h, f''_z h \rangle \geq 0$, and $\nabla f(x) \in E^*$, this implies $f(x) \leq \liminf_{n \to \infty} f(x_n)$. Q.E.D.

We can deduce growth conditions in a similar manner.

THEOREM 1.4.4. Let f have first and second derivatives throughout E and let f''_x satisfy at each $x \in E$ the condition $\langle h, f''_x h \rangle \geq ||h|| g(||h||)$ where g is a nonnegative continuous function for $t \geq 0$ tending to infinity with t. Then $\lim_{||x|| \to \infty} f(x) = +\infty$.

Proof: Let $s(t) = \langle x, \nabla f(tx) \rangle$ for $0 \leq t \leq 1$; then $s(1) = s(0) + \int_0^1 s'(t)\, dt$, that is,

$$\langle x, \nabla f(x) \rangle = \langle x, \nabla f(0) \rangle + \int_0^1 \langle x, f''_{tx} x \rangle\, dt$$

Therefore, $\langle x, \nabla f(x) \rangle \geq -||x||\,||\nabla f(0)|| + ||x|| g(||x||)$. Since

$$\begin{aligned} f(x) &= f(0) + \int_0^1 \langle x, \nabla f(tx) \rangle\, dt \\ &= f(0) + \int_0^1 \langle tx, \nabla f(tx) \rangle \frac{1}{t}\, dt \\ &\geq f(0) - ||x||\,||\nabla f(0)|| + ||x|| \int_0^1 g(t||x||)\, dt \\ &\geq f(0) + ||x|| \left[-||\nabla f(0)|| + \int_0^1 g(t||x||)\, dt \right] \end{aligned}$$

and since $\lim_{t \to \infty} g(t) = \infty$, we then conclude that $\lim_{||x|| \to \infty} \int_0^1 g(t||x||)\, dt = \infty$. Q.E.D.

EXERCISE. Consider

$$f(x) = \int_0^1 \{\tfrac{1}{2}[\dot{x}(t)]^2 + c[t, x(t)]\}\, dt, \qquad \dot{x}(t) \equiv \frac{d}{dt} x(t)$$

for x in

$$E = {}_0W_2^1(0, 1) \equiv \{x; x \in W_2^1(0, 1), x(0) = x(1) = 0\}$$

(See Example 3, Section 1.3.) We let $||x||^2 = \int_0^1 [\dot{x}(t)]^2\, dt$. Suppose

$$c_{uu}(t, u) \equiv \frac{\partial^2 c(t, u)}{\partial u^2} \geq \gamma > -\pi^2$$

for all t, u. Prove that $\lim_{||x|| \to \infty} f(x) = +\infty$. *Hint:* Use the fact that for all $x \in E$ we have $\int_0^1 [\dot{x}(t)]^2 \, dt \geq \pi^2 \int_0^1 [x(t)]^2 \, dt$ to prove that $\langle f''_x h, h \rangle \geq \epsilon \langle h, h \rangle$, for small enough $\epsilon > 0$.

As we stated earlier, we wish to see how minimization problems are related to the solution of operator equations. The fundamental statement is as follows.

THEOREM 1.4.5. Let the functional f defined on a set S in a Banach space E be minimized at a point $x_0 \in S$, with x_0 an interior point in the norm topology. If f has a derivative $\nabla f(x_0)$ at x_0, then $\nabla f(x_0) = 0$.

Proof: For any fixed $h \in E$, $f(x_0 + th)$ is a real-valued function of the real variable t on some open interval containing zero, having a derivative at zero, and being minimized there. Therefore,

$$0 = \frac{d}{dt} f(x_0 + th)\bigg|_{t=0} = \langle h, \nabla f(x_0) \rangle$$

Since $h \in E$ was arbitrary, $\langle h, \nabla f(x_0) \rangle = 0$ for all $h \in E$ and, therefore, $\nabla f(x_0) = 0$. Q.E.D.

Similarly, minimization problems over certain types of sets can be related to *generalized eigenvalue* problems via Lagrange multiplier theorems. As an example, we merely state one type of such result.

PROPOSITION 1.4.2. Let f and g be functionals on a Banach space E, both of which are differentiable at the point x_0 which is local minimum of f on the set $\{x; g(x) = g(x_0)\}$. If $||\nabla g(x_0)|| > 0$, then there exists a real number λ such that $\nabla f(x_0) = \lambda \nabla g(x_0)$.

Proposition 1.4.2 gives a necessary condition for a point to give a local minimum over a set defined by one real equality constraint. Much of the theory of optimal control and mathematical programming involves determining necessary conditions and also additional sufficient conditions for a point to be an extremizing point over sets with more complicated constraints. Another simple necessary condition for x_0 to minimize a differentiable $f(x)$ over a convex set C is that $\langle x - x_0, \nabla f(x_0) \rangle \geq 0$ for all x in C. The amount of literature in this extremely complex area is vast. Since our main concern here is with the theory of and methods for approximate solutions primarily to unconstrained problems, we shall pursue this no further.

General references: Levitin–Poljak (1966a), Vainberg (1964).

1.5. THE ROLE OF CONVEXITY

We saw already in Corollary 1.4.1 that convexity of the set over which we seek a minimum can be of importance in proving existence of a solution; various kinds of convexity assumptions about the functional to be minimized also play a vital role. We shall consider next some examples of how these properties are related to continuity properties of the functional and to the nature of the points solving the minimization problem. Although more general definitions are possible, for the following let C be a convex subset of a Banach space E.

DEFINITION 1.5.1. A functional f is *convex* (*strictly convex*) on C if and only if for each x_1, x_2 in C and λ in $(0, 1)$,

$$f[\lambda x_1 + (1 - \lambda)x_2] \leq (<) \, \lambda f(x_1) + (1 - \lambda)f(x_2)$$

EXERCISE. When is Definition 1.5.1 equivalent to the following?

$$f\left(\frac{x_1 + x_2}{2}\right) \leq \frac{1}{2}f(x_1) + \frac{1}{2}f(x_2)$$

This definition is the natural extension of the concept of a convex real-valued function of a real variable to that of a functional on a general space. The usual characterizations for differentiable convex functionals are valid here also; we state them without proof.

PROPOSITION 1.5.1. A Frechet-differentiable functional f is convex (strictly convex) on a norm-open convex set C if and only if

$$f(x_2) - f(x_1) \geq (>) \, \langle x_2 - x_1, \nabla f(x_1) \rangle$$

for each x_1, x_2 in C, $x_1 \neq x_2$. Another equivalent condition is that

$$\langle x_2 - x_1, \nabla f(x_2) - \nabla f(x_1) \rangle \geq (>) \, 0$$

for x_1, x_2 in C, $x_1 \neq x_2$. If f''_x exists in the convex open set C, then f is convex if and only if $\langle h, f''_x h \rangle \geq 0$ for all h in E and x in C.

If we combine the last statement in Proposition 1.5.1 with Theorem 1.4.3, we immediately see that a twice-differentiable convex functional on an open convex set C is weakly sequentially lower semicontinuous on C. The hypotheses leading to this conclusion can be weakened considerably, leading to a generalized concept of convexity which, as we shall see, is in a certain sense the natural hypothesis for minimization problems. First we observe that if f is convex on E, then $\{x; f(x) \leq \alpha\}$ is convex for each fixed real α; we take this as a defining property.

DEFINITION 1.5.2. A functional f is *quasi-convex* on the convex set C if and only if $\{x; f(x) \leq \alpha, x \in C\}$ is convex for each fixed real α. This is equivalent to

$$f[\lambda x_1 + (1 - \lambda)x_2] \leq \max [f(x_1), f(x_2)\}$$

for all λ in (0, 1), x_1, x_2 in C.

EXERCISE. Prove that the two statements in Definition 1.5.2 are equivalent.

EXERCISE. When is Definition 1.5.2 equivalent to the following?

$$f\left(\frac{x_1 + x_2}{2}\right) \leq \max \{f(x_1), f(x_2)\}$$

Thus a convex functional is quasi-convex. So also is the product of two nonnegative convex functions [Tremolieres (1969)] or the quotient of a convex, nonpositive function by a convex, positive function [Mangasarian (1969)].

EXERCISE. Prove that a convex functional is quasi-convex.

We can now prove a weakened form of our earlier statement concerning the weak lower semicontinuity of a twice-differentiable convex functional on an open convex set.

THEOREM 1.5.1. Let f be a quasi-convex, norm sequentially lower semicontinuous functional defined on a convex set C. Then f is weakly sequentially lower semicontinuous on C.

Proof: Let $x_n \rightharpoonup x$, $x_n \in C$, $x \in C$, and let x_{n_i} be a subsequence such that $\lim_{i\to\infty} f(x_{n_i}) = \liminf_{n\to\infty} f(x_n)$. According to a statement in Section 1.3, there exist for each l convex combinations $y_{l,i}$ of the points x_{n_i} for $i \geq l$ norm converging to x, that is,

$$y_{l,i} \equiv \sum_{j=l}^{i} \lambda_{l,j} x_{n_j}$$

$$0 \leq \lambda_{l,j} \leq 1$$

$$\sum_{j=l}^{i} \lambda_{l,j} = 1$$

and

$$\lim_{i\to\infty} ||y_{l,i} - x|| = 0$$

Then $f(x) \leq \liminf_{i\to\infty} f(y_{l,i})$. But by the quasi-convexity one can see that $f(y_{l,i}) \leq \max_{l \leq j \leq i} f(x_{n_j})$. Thus

$$f(x) \leq \liminf_{i \to \infty} f(y_{l,i}) \leq \sup_{j \geq l} f(x_{n_j})$$

Letting l tend to infinity we find

$$f(x) \leq \limsup_{j \to \infty} f(x_{n_j}) = \lim_{j \to \infty} f(x_{n_j}) = \liminf_{n \to \infty} f(x_n)$$

Q.E.D.

EXERCISE. Prove that a quasi-convex, norm lower semicontinuous functional defined on a weakly closed convex set C is weakly lower semicontinuous on C by considering $\{x, f(x) \leq \alpha\}$ for real α.

Thus a functional as described by Theorem 1.5.1 achieves its minimum on each weakly sequentially compact set; it is simple to describe hypotheses under which the minimizing point is unique.

DEFINITION 1.5.3. A functional f is called *strongly quasi-convex* on a convex set C if and only if for $x_1 \neq x_2$,

$$f[\lambda x_1 + (1 - \lambda)x_2] < \max \{f(x_1), f(x_2)\}$$

for λ in (0, 1).

EXERCISE. When is Definition 1.5.3 equivalent to the following?

$$f\left(\frac{x_1 + x_2}{2}\right) < \max \{f(x_1), f(x_2)\}$$

From strong quasi-convexity follows a uniqueness theorem:

THEOREM 1.5.2. A strongly quasi-convex functional can achieve its minimum over a convex set C at no more than one point.

Proof: If $f(x_1) = f(x_2) = \inf_{x \in C} f(x)$, then $\frac{1}{2}x_1 + \frac{1}{2}x_2 \in C$, but

$$f(\tfrac{1}{2}x_1 + \tfrac{1}{2}x_2) < \max \{f(x_1), f(x_2)\} = \inf_{x \in C} f(x)$$

which is a contradiction. Q.E.D.

In a sense, strong quasi-convexity is the natural assumption for minimization problems. More precisely, it is easy to show that if f is a norm-continuous functional on a reflexive space E, then it achieves its minimum at a unique point over each norm-closed, convex, bounded set if and only if f is strongly quasi-convex [Poljak (1966)].

EXERCISE. Prove that a norm-continuous functional f defined in a reflexive space achieves its minimum at a unique point over each norm-closed, convex, bounded set if and only if f is strongly quasi-convex.

A related but weaker concept is that of *strict quasi-convexity*, in which the inequality in Definition 1.5.3 is required to hold only if $f(x_1) \neq f(x_2)$; strongly quasi-convex functionals and convex functionals are strictly quasi-convex, but a strictly quasi-convex functional need not even be quasi-convex.

EXERCISE. Prove that both strongly quasi-convex functionals and convex functionals are strictly quasi-convex; find a strictly quasi-convex functional that is not quasi-convex.

The importance of strict quasi-convexity can be seen from the following theorem.

THEOREM 1.5.3. Let x_0 yield a local minimum for the strictly quasi-convex functional f over the convex set C. Then x_0 yields the global minimum.

Proof: If there exists an x_1 in C with $f(x_1) < f(x_0)$, then $f(\lambda x_0 + (1 - \lambda)x_1) < f(x_0)$ for all λ in $(0, 1)$ and λ arbitrarily close to 1. But then x_0 cannot yield a local minimum. Q.E.D.

Convexity assumptions often appear in somewhat different form than discussed so far. For example, in the calculus-of-variations problem of minimizing

$$f(x) = \int_0^1 c[t, x(t), \dot{x}(t)]\, dt$$

with certain boundary conditions, a standard assumption [Akhiezer (1962)] is that $c(t, x, y)$ is differentiable and convex in the variable y for fixed (t, x), a condition to which we shall return in Section 3.5; under some additional continuity and growth hypotheses on $c(t, x, y)$, one can deduce the weak lower semicontinuity of f in the appropriate space. This is really a special case of the following theorem generalizing Theorem 1.5.1 [Nashed (1967)].

THEOREM 1.5.4. Let the functional $c(x, y)$ on the space $E \times E$ for a Banach space E be norm lower semicontinuous and quasi-convex in y for each fixed x, and weakly sequentially lower semicontinuous in x on bounded sets, uniformly so for y in a bounded set. Then $f(x) \equiv c(x, x)$ is weakly sequentially lower semicontinuous.

Proof: Let $x_n \rightharpoonup x'$. For any subsequence $\{x_{n_i}\}$, if $c(x', x_{n_i})$ converges to r, then for all $\lambda > r$ we consider

$$C_\lambda = \{x;\, c(x', x) \leq \lambda\}$$

which by the norm lower semicontinuity and quasi-convexity of $c(x, y)$ in y must be norm-closed and convex, and hence weakly closed. Thus $x' \in C_\lambda$ for all $\lambda > r$, so

$$c(x', x') \leq \liminf_{n\to\infty} c(x', x_n)$$

Now,

$$f(x') = c(x', x') = c(x', x') - c(x', x_n) + c(x', x_n) - c(x_n, x_n) + f(x_n)$$

From the preceding inequality and the uniform (in y) weak sequential lower semicontinuity of $c(x, y)$ in x, it follows that

$$f(x') \leq \liminf_{n\to\infty} f(x_n)$$

Q.E.D.

Convexity properties are also of vital importance from a more computational viewpoint; we consider this aspect next.

General references: Levitin–Poljak (1966a, b), Mangasarian (1969), Ponstein (1967).

1.6. MINIMIZING SEQUENCES

In computing a sequence of approximate solutions x_n to a minimization problem, one of our goals nearly always is to be sure that $f(x_n)$ is converging to the minimum value of f; thus we desire $\{x_n\}$ to be a *minimizing sequence*. Often other conclusions, such as convergence of $\{x_n\}$, follow directly from this fact. For example, if $\{x_n\}$ is a minimizing sequence of elements from the weakly sequentially compact set C for the weakly sequentially lower semicontinuous f over C, then for any weakly convergent subsequence (at least one of which of course exists) $\{x_{n_i}\}$ converging to an x', we have

$$f(x') \leq \liminf_{i\to\infty} f(x_{n_i}) = \inf_{x\in C} f(x)$$

that is, x' minimizes f over C; if the minimizing point is unique, then $\{x_n\}$ weakly converges to it. The above argument is extremely common and is almost always the least we can say about convergence in some scheme of approximate minimization; often it is possible to say much more—for example, to deduce norm convergence. We indicate next some of the typical concepts that are of value in this respect; in most of the rest of this book we shall not state all the various conclusions about stronger convergence in each approximation method which can be drawn from the type of considerations to be presented now, but leave such synthesizing to the student.

We now take a more general view. Since in many practical cases the approximate solutions can be guaranteed only approximately to satisfy the constraints on the problem, we introduce the concept of an approximate minimizing sequence (or generalized minimizing sequence [Levitin–Poljak (1966a, b), Poljak (1966)]).

DEFINITION 1.6.1. The sequence $\{x_n\}$ is an *approximate minimizing sequence* for the functional f over the set C if $\lim_{n\to\infty} d(x_n, C) = 0$, where $d(x_n, C) = \inf_{x \in C} ||x_n - x||$, and $\lim_{n\to\infty} f(x_n) = \inf_{x \in C} f(x)$.

PROPOSITION 1.6.1. If f is a weakly sequentially lower semicontinuous functional on a norm neighborhood of the weakly sequentially compact set C and $\{x_n\}$ is an approximate minimizing sequence, then all weak limit points of $\{x_n\}$, at least one of which exists, lie in C and minimize f over C. If f has a unique minimizing point x^* in C, then $x_n \rightharpoonup x^*$.

EXERCISE. Prove Proposition 1.6.1 above.

The goal of this section is to discover conditions under which the convergence is more useful—that is, so that $\{x_n\}$ itself converges or so that we have norm convergence. We saw in Section 1.5 that strong quasi-convexity is in a sense the natural hypothesis for minimization problems; a stronger form of it will allow us to deduce norm convergence of approximate minimizing sequences.

DEFINITION 1.6.2. A quasi-convex functional is called *uniformly quasi-convex* on a convex set C if and only if there is a real-valued, continuous, monotone-increasing function $\delta(t)$ for $t \geq 0$ with $\delta(t) = 0$ if and only if $t = 0$, such that

$$f\left(\frac{x+y}{2}\right) \leq \max\{f(x), f(y)\} - \delta(||x - y||)$$

for all x, y in C.

EXERCISE. Find a condition equivalent to the one in Definition 1.6.2 for $f[\lambda x + (1 - \lambda)y]$, λ in $[0, 1]$.

THEOREM 1.6.1. Let C be a norm-closed, bounded, convex subset of a reflexive space E; f a weakly sequentially lower semicontinuous uniformly quasi-convex functional on a norm neighborhood of C; and $\{x_n\}$ an approximate minimizing sequence. Then $x_n \to x^*$, the unique point in C minimizing f.

Proof: By Proposition 1.6.1, there exists at least one subsequence x_{n_i} weakly converging to a minimizing point x'. But the minimizing point $x' = x^*$ clearly is unique, just as in Theorem 1.5.2, so $x_n \rightharpoonup x^*$. Since $f(x_n)$ converges to $f(x^*)$, since $(x_n + x^*)/2 \rightharpoonup x^*$, and since f is weakly sequentially lower semicontinuous, for all $\epsilon > 0$ there is an N_ϵ such that $n \geq N_\epsilon$ implies

$$|f(x_n) - f(x^*)| < \epsilon \qquad \text{and} \qquad f\left(\frac{x_n + x^*}{2}\right) > f(x^*) - \epsilon$$

By the uniform quasi-convexity,

$$\delta(||x_n - x^*||) \leq \max\{f(x_n), f(x^*)\} - f\left(\frac{x_n + x^*}{2}\right) < 2\epsilon \quad \text{for} \quad n \geq N_\epsilon$$

and hence $||x_n - x^*|| \longrightarrow 0$. Q.E.D.

It is also possible to prove strong convergence results on the basis of properties of C, not f.

DEFINITION 1.6.3. A set C is called *uniformly convex* if and only if there exists a real-valued, continuous, monotone function $\delta(t)$ for $t \geq 0$ with $\delta(t) = 0$ if and only if $t = 0$, such that $x, y \in C$ and $||z|| \leq \delta(||x - y||)$ imply $(x + y)/2 + z \in C$.

THEOREM 1.6.2. Let C be a norm-closed, bounded, uniformly convex subset of a reflexive space E; f a weakly sequentially lower semicontinuous functional on a norm neighborhood of C with a unique point x^* minimizing f on C; x^* on the norm boundary of C; and $\{x_n\}$ an approximate minimizing sequence. Then $x_n \longrightarrow x^*$.

Proof: By Proposition 1.6.1 and the uniqueness of x^*, we have $x_n \rightharpoonup x^*$. Since $d(x_n, C) \longrightarrow 0$, there is a sequence $x'_n \in C$ with $\lim_{n\to\infty} ||x_n - x'_n|| = 0$. Suppose there exists $\epsilon > 0$ and a subsequence $\{x'_{n_i}\}$ with $||x'_{n_i} - x^*|| \geq \epsilon$ for all i. By the uniform convexity of C, if $||z|| \leq \delta(\epsilon)$, then

$$\frac{x'_{n_i} + x^*}{2} + z \in C$$

for all i. Since

$$\frac{x'_{n_i} + x^*}{2} + z \rightharpoonup x^* + z$$

and C is weakly closed, $x^* + z \in C$ for all z satisfying $||z|| \leq \delta(\epsilon) > 0$. Then x^* is not a boundary point, yielding a contradiction. Thus x'_n (and hence also x_n) converges in norm to x^*. Q.E.D.

EXERCISE. Consider $f(x) = \int_0^1 \{t[\dot{x}(t)]^2 + [x(t)]^2\}\,dt$ on ${}_0W_2^1(0, 1)$, the set of x in $W_2^1(0, 1)$ with $x(0) = x(1) = 0$ and with $||x||^2 \equiv \int_0^1 [\dot{x}(t)]^2\,dt$. Let

$$x_n(t) \equiv \begin{cases} n^{1/2}t \quad \text{for} \quad 0 \leq t \leq \dfrac{1}{n} \\ n^{-1/2} - n^{1/2}\left(t - \dfrac{1}{n}\right) \quad \text{for} \quad \dfrac{1}{n} \leq t \leq \dfrac{2}{n} \\ 0 \quad \text{for} \quad \dfrac{2}{n} \leq t \leq 1 \end{cases}$$

Show that $\{x_n\}$ is a weakly convergent but *not* norm-convergent minimizing sequence. Is there a better norm to use for this problem?

Computationally one often desires to have some type of error bound or estimate of the rate of convergence; such bounds can be found for certain classes of problems. As an example, we state merely one simple theorem.

THEOREM 1.6.3. Let C be a norm-closed, bounded, convex subset of a reflexive space E; f a twice-differentiable functional whose second derivative satisfies $\langle h, f''_x h\rangle \geq \delta(||h||)$ for all x in a norm neighborhood of C where $\delta(t)$ is a real-valued, continuous, monotone function for $t \geq 0$ with $\delta(t) = 0$ if and only if $t = 0$; and $\{x_n\}$ an approximate minimizing sequence. Then $x_n \longrightarrow x^*$, the unique point minimizing f in C, and

$$\delta(||x_n - x^*||) \leq f(x_n) - f(x^*)$$

Proof: According to Proposition 1.5.1, f is convex. It is also uniformly quasi-convex, since the equations

$$f(x) = f\left(\frac{x+y}{2} + \frac{x-y}{2}\right) \geq f\left(\frac{x+y}{2}\right) + \left\langle \frac{x-y}{2}, \nabla f\left(\frac{x+y}{2}\right)\right\rangle + \frac{1}{2}\delta\left(\frac{||x-y||}{2}\right)$$

$$f(y) = f\left(\frac{x+y}{2} - \frac{y-x}{2}\right) \geq f\left(\frac{x+y}{2}\right) - \left\langle \frac{x-y}{2}, \nabla f\left(\frac{x+y}{2}\right)\right\rangle + \frac{1}{2}\delta\left(\frac{||x-y||}{2}\right)$$

imply

$$f\left(\frac{x+y}{2}\right) \leq \frac{1}{2}f(x) + \frac{1}{2}f(y) - \frac{1}{2}\delta\left(\frac{||x-y||}{2}\right)$$

a stronger condition (*uniform convexity*) than asserted. Thus by Theorem 1.6.1, this theorem is proved except for the error estimate. In the remarks following Proposition 1.4.2 we noted that we must have $\langle x - x^*, \nabla f(x^*)\rangle \geq 0$ for all x in C; otherwise, f decreases in the direction of x, a contradiction to the optimality of x^*. Hence

$$f(x) - f(x^*) \geq \langle x - x^*, \nabla f(x^*)\rangle + \delta(||x - x^*||) \geq \delta(||x - x^*||)$$

Q.E.D.

We shall have cause to make use of the type of estimate in the above proof later, especially when f''_x is a *uniformly positive-definite* self-adjoint operator in a Hilbert space, that is, $\delta(t) = mt^2$, $m > 0$, so that

$$||x_n - x^*|| \leq \left\{\frac{f(x_n) - f(x^*)}{m}\right\}^{1/2}$$

In order for bounds of the form $\delta(||x_n - x^*||) \leq f(x_n) - f(x^*)$ to be computable, we need a good computable lower bound—say m_n—for $f(x^*)$, so that we have $\delta(||x_n - x^*||) \leq f(x_n) - m_n$; we now indicate how this can be done in some cases by *complementary variational principles*, which have been rather thoroughly studied for differential equations [Noble (1964, 1966)], operator-theory extensions thereof [Rall (1966)], and error estimation for variational boundary-value problems [Shampine (1968)].

THEOREM 1.6.4. Let f be a real-valued nonlinear functional on a real Hilbert space H and let $\hat{u}$ minimize f on the norm-closed convex set S. For each w in some given set $S^* \subset H$, let a self-adjoint linear operator P_w be defined, satisfying $\langle P_w h, h\rangle \geq a_w\langle h, h\rangle$ for all h in H, with $a_w > 0$. Suppose that f is twice continuously Frechet-differentiable on the set of points of the form $\lambda v + (1 - \lambda)w$ for v in S, w in S^*, and λ in [0, 1]; and that

$$\langle [f''_{\lambda v+(1-\lambda)w} - P_w](v - w), v - w\rangle \geq 0$$

for all v in S, w in S^*, and λ in [0, 1]. For each w in S^*, let v_w in S be defined as the (unique) point in S minimizing $\frac{1}{2}\langle P_w v, v\rangle + \langle \nabla f(w) - P_w w, v\rangle$, and let f^* be defined on S^* as

$$f^*(w) = f(w) + \tfrac{1}{2}\langle v_w - w, P_w(v_w - w)\rangle + \langle v_w - w, \nabla f(w)\rangle$$

If $\hat{u}$ is in S^*, then $f^*(w)$ is maximized by $w = \hat{u}$, and $f^*(\hat{u}) = f(\hat{u})$. For all u in S and w in S^*, we have the estimates

$$f(u) - f^*(w) \geq \tfrac{1}{2}\langle P_w(\hat{u} - v_w), \hat{u} - v_w\rangle$$
$$f(u) - f^*(w) \geq \tfrac{1}{2}\langle P_{\hat{u}}(\hat{u} - u), \hat{u} - u\rangle$$

Proof: Since $\hat{u}$ minimizes f over the convex set S, we have $\langle \nabla f(\hat{u}), s - \hat{u}\rangle \geq 0$ for all s in S. Let

$$g(v) = \tfrac{1}{2}\langle P_{\hat{u}} v, v\rangle + \langle \nabla f(\hat{u}) - P_{\hat{u}}\hat{u}, v\rangle$$

Then

$$\begin{aligned}\langle s - \hat{u}, \nabla g(\hat{u})\rangle &= \langle s - \hat{u}, P_{\hat{u}}\hat{u} + \nabla f(\hat{u}) - P_{\hat{u}}\hat{u}\rangle \\ &= \langle s - \hat{u}, \nabla f(\hat{u})\rangle \geq 0\end{aligned}$$

for all s in $\hat{u}$. Since the quadratic functional $g(v)$ has $g''_v \equiv P_{\hat{u}}$, a positive-definite operator, a unique point $v_{\hat{u}}$ exists (similarly for v_w for all w in S^*) and is characterized by the condition $\langle s - v_{\hat{u}}, \nabla g(v_{\hat{u}})\rangle \geq 0$; therefore, $v_{\hat{u}} = \hat{u}$ and hence $f^*(\hat{u}) = f(\hat{u})$. To prove that $\hat{u}$ maximizes f^*, we write, for w in S^*, letting $v_w \equiv v$ and $P_w \equiv P$ for notational ease,

$$\begin{aligned} f^*(\hat{u}) - f^*(w) &= f(\hat{u}) - f(w) - \tfrac{1}{2}\langle v - w, P(v - w)\rangle - \langle v - w, \nabla f(w)\rangle \\ &= \tfrac{1}{2}\langle P(\hat{u} - v), \hat{u} - v\rangle + [f(\hat{u}) - \tfrac{1}{2}\langle P\hat{u}, \hat{u}\rangle] \\ &\quad - [f(w) - \tfrac{1}{2}\langle Pw, w\rangle] + \langle \hat{u} - v, Pv - Pw + \nabla f(w)\rangle \\ &\quad + \langle \hat{u} - w, Pw - \nabla f(w)\rangle \end{aligned}$$

by adding and subtracting $\frac{1}{2}\langle P(\hat{u} - v), \hat{u} - v\rangle$. Since v minimizes $\frac{1}{2}\langle Pv, v\rangle + \langle \nabla f(w) - Pw, v\rangle$ over S, and $\hat{u}$ is in S, the inner-product inequality characterizing v gives $\langle \hat{u} - v, Pv - Pw + \nabla f(w)\rangle \geq 0$. Hence

$$\begin{aligned} f^*(\hat{u}) - f^*(w) \geq \tfrac{1}{2}\langle P(\hat{u} - v), \hat{u} - v\rangle + d(\hat{u}) - d(w) \\ + \langle \hat{u} - w, Pw - \nabla f(w)\rangle \end{aligned}$$

where $d(u) \equiv f(u) - \frac{1}{2}\langle Pu, u\rangle$. We have

$$\begin{aligned} d(\hat{u}) - d(w) &= \langle \nabla d(w), \hat{u} - w\rangle + \int_0^1 t\langle d''_{t\hat{u}+(1-t)w}(\hat{u} - w), \hat{u} - w\rangle\, dt \\ &= \langle \nabla f(w) - Pw, \hat{u} - w\rangle \\ &\quad + \int_0^1 t\langle [f''_{t\hat{u}+(1-t)w} - P_w](\hat{u} - w), \hat{u} - w\rangle\, dt \\ &\geq \langle \nabla f(w) - Pw, \hat{u} - w\rangle \end{aligned}$$

Inserting this we find

$$\begin{aligned} f^*(\hat{u}) - f^*(w) &\geq \tfrac{1}{2}\langle P(\hat{u} - v), \hat{u} - v\rangle + \langle \nabla f(w) - Pw, \hat{u} - w\rangle \\ &\quad + \langle \hat{u} - w, Pw - \nabla f(w)\rangle \\ &\geq \frac{1}{2}\langle P(\hat{u} - v), \hat{u} - v\rangle \geq \frac{a_w}{2}||\hat{u} - v||^2 \end{aligned}$$

Thus $\hat{u}$ maximizes f^* over S^*. To obtain the first error estimate, we merely write

$$f(u) - f^*(w) \geq f(\hat{u}) - f^*(w) = f^*(\hat{u}) - f^*(w) \geq \tfrac{1}{2}\langle P_w(\hat{u} - v_w), \hat{u} - v_w\rangle$$

For the second, we write

$$\begin{aligned} f(u) - f^*(w) \geq f(u) - f^*(\hat{u}) &= f(u) - f(\hat{u}) = \langle u - \hat{u}, \nabla f(\hat{u})\rangle \\ &\quad + \int_0^1 t\langle [f''_{tu+(1-t)\hat{u}} - P_{\hat{u}}](u - \hat{u}), u - \hat{u}\rangle\, dt \\ &\quad + \tfrac{1}{2}\langle P_{\hat{u}}(u - \hat{u}), u - \hat{u}\rangle \\ &\geq \tfrac{1}{2}\langle P_{\hat{u}}(u - \hat{u}), u - \hat{u}\rangle \end{aligned}$$

by the assumption on $f'' - P$ and the necessary condition for $\hat{u}$ to minimize f over S, since u is in S.

Remarks on Theorem 1.6.4.

1. The closedness and convexity of S are used only to guarantee the existence and uniqueness of v_w for w in S and to deduce that $v_{\hat{u}} = \hat{u}$; these properties can be guaranteed in other ways as well.

2. The differentiability hypotheses can be weakened easily; in particular, $f(v) - \frac{1}{2}\langle P_w v, v\rangle$ need only be differentiably convex on S for each w in S^*.

3. Independent of any convexity hypotheses on f, points other than $\hat{u}$ can maximize f^* without added restrictions on $f'' - P$; this is of no consequence for our purposes of error-bounding, however.

4. In order for the error bounds to be effective, one would require that f^* be continuous at $\hat{u}$; this requires further study of P_w. An examination of the expression for $f^*(\hat{u}) - f^*(w)$ shows that if f''_u and P_w are uniformly bounded for u and w near $\hat{u}$, then for some constants a, b, c we have

$$0 \leq f^*(\hat{u}) - f^*(w) \leq a||\hat{u} - v_w||^2 + b||\hat{u} - w||^2 + c||\hat{u} - v_w|| \, ||\hat{u} - w||$$

Thus we need only study $\hat{u} - v_w = v_{\hat{u}} - v_w$ for w near $\hat{u}$. If, for example, $a_w \geq \epsilon > 0$, $||\nabla f(w) - \nabla f(\hat{u})|| \leq K||w - \hat{u}||$ near $\hat{u}$, and $S = H$, then $||\hat{u} - v_w|| \leq [1 + (K/\epsilon)]||\hat{u} - w||$.

For the general situation we have the following more restrictive results:

THEOREM 1.6.5. Let the assumptions of Theorem 1.6.4 hold. Moreover, suppose that $||f''_w|| \leq A$, $||P_w - P_{\hat{u}}|| \leq A||w - \hat{u}||$, and $a_w \geq \epsilon > 0$, all for w near $\hat{u}$. Then $||\hat{u} - v_w|| = O(||\hat{u} - w||^{1/2})$ and hence $f^*(\hat{u}) - f^*(w) = O(||\hat{u} - w||)$.

Proof: Let

$$g_w(v) \equiv \tfrac{1}{2}\langle P_w v, v\rangle + \langle \nabla f(w) - P_w w, v\rangle$$

Then

$$|g_{\hat{u}}(v) - g_w(v)| = O(||v|| \, ||\hat{u} - w||)$$

Then we have

$$g_w(v_w) \leq g_w(\hat{u}) = g_{\hat{u}}(\hat{u}) + [g_w(\hat{u}) - g_{\hat{u}}(\hat{u})] \leq g_{\hat{u}}(\hat{u}) + O(||\hat{u}|| \, ||\hat{u} - w||)$$

and, similarly,

$$g_{\hat{u}}(\hat{u}) \leq g_{\hat{u}}(v_w) \leq g_w(v_w) + O(||v_w|| \, ||\hat{u} - w||)$$

Hence

$$|g_{\hat{u}}(\hat{u}) - g_w(v_w)| \leq ||v_w|| O(||\hat{u} - w||)$$

Now also for w near $\hat{u}$, $g_w(v) \geq \frac{1}{2}a_w\|v\|^2 - \|v\|M$ for a fixed constant M; and since $g_w(0) = 0$, we have $\|v_w\| \leq 2M/\epsilon$. In addition,

$$g_{\hat{u}}(v) = g_{\hat{u}}(\hat{u}) + \langle \nabla g_{\hat{u}}(\hat{u}), v - \hat{u} \rangle + \frac{1}{2}\langle P_{\hat{u}}(v - \hat{u}), v - \hat{u} \rangle$$

$$\geq g_u(\hat{u}) + \frac{\epsilon}{2}\|v - \hat{u}\|^2$$

Therefore,

$$\frac{\epsilon}{2}\|v_w - \hat{u}\|^2 \leq g_{\hat{u}}(v_w) - g_{\hat{u}}(\hat{u}) \leq g_w(v_w) + O(\|v_w\|\|\hat{u} - w\|) - g_{\hat{u}}(\hat{u})$$
$$\leq O(\|\hat{u} - w\|) + g_w(v_w) - g_{\hat{u}}(\hat{u}) \leq O(\|\hat{u} - w\|)$$

Q.E.D.

We wish to give two concrete examples illustrating the meaning of the general theorem above, Theorem 1.6.5; it is simplest to consider differential equations, and in order to minimize technical complexities we consider the equation

$$u''(t) = c[t, u(t)], \qquad t \text{ in } (0, 1), \qquad u(0) = u(1) = 0$$

More precisely, we consider minimizing the functional

$$f(u) \equiv \tfrac{1}{2}\int_0^1 [u'(t)]^2\, dt + \int_0^1 \int_0^{u(t)} c(t, x)\, dx\, dt$$

over the set $H \equiv {}_0W_2^1(0, 1)$. For u, v in H, we take

$$\langle u, v \rangle = \int_0^1 [u'(t)v'(t) + u(t)v(t)]\, dt$$

Since we wish to illustrate *ideas* rather than *technicalities*, we shall be rather sloppy and speak blithely of $D^2u \equiv u''$ for u in H; the precise formulation is easily filled in.

A. For the first example, let us suppose that $c_u(t, u) \geq \gamma > -\pi^2$ for all u in $(-\infty, \infty)$, t in $[0, 1]$. Let $(u, v) = \int_0^1 u(t)v(t)\, dt$. Then

$$f(u + h) = f(u) + (-D^2u + c(t, u), h)$$
$$+ \tfrac{1}{2}([-D^2 + c_u(t, u)]h, h) + \text{small terms}$$

Thus

$$\langle \nabla f(u), h \rangle = (-D^2u + c(t, u), h)$$

and

$$\langle f_u''h, h\rangle = ([-D^2 + c_u(t, u)]h, h)$$

We let $S = S^* = H$ and for all w define P_w by $-D^2 + \gamma$ which is positive-definite since $\gamma > -\pi^2$. Since $c_u \geq \gamma$, we have

$$\langle [f_u'' - P_w]h, h\rangle = ([c_u - \gamma]h, h) \geq 0$$

and the hypotheses are fulfilled. Here

$$\begin{aligned} f^*(w) = {} & \frac{1}{2}\int_0^1 [w'(t)]^2\, dt \\ & + \int_0^1 \Big\{\frac{\gamma}{2}v^2(t) + [v(t) - w(t)]\{c[t, w(t)] - \gamma w(t)\} \\ & + \int_0^{w(t)} [c(t, x) - \gamma x]\, dx\Big\}\, dt \end{aligned}$$

where $v(t)$ solves

$$v''(t) - \gamma v(t) = c[t, w(t)] - \gamma w(t) \quad \text{for} \quad t \text{ in } (0, 1), \qquad v(0) = v(1) = 0$$

In this case, discussed in Shampine (1968), the error bounds are in the norm

$$\langle Pe, e\rangle = \int_0^1 \{[e'(t)]^2 + \gamma e^2(t)\}\, dt$$

This yields useful bounds for any approximate solution w and for the corresponding v_w. Such a w might, for example, be obtained by the Ritz procedure or by an iterative process. For some problems the Newton iterative process yields a sequence u_n decreasing to the desired solution [Bellman (1957, 1962), Collatz (1966), Shampine (1966)]; often, then v_{u_n} turns out to lie below the solution [Shampine (1966)] yielding error bounds for $\hat{u}$. The variational procedure above in addition furnishes bounds involving the derivatives.

B. In some cases, the Newton iteration mentioned above may be costly to carry out. Certain Picard-type iterations, though more slowly convergent, are sometimes used at least until one is near the solution where Newton's method might be worth the cost. The process above in A will often yield two-sided bounds and the $\langle P_e e, e\rangle$ bounds as well in this case too. We wish to observe that one Newton step also provides such bounds in some cases. Suppose now that $|c(t, u)| \leq N$ for all t, u; that u_0 solves $u_0'' = -N$, $u_0(0) = u_0(1) = 0$; and that $k \geq c_u(t, u) \geq \gamma > -\pi^2$. Then it is known that the sequence

$$u_{n+1}'' - ku_{n+1} = c(t, u_n) - ku_n, \qquad u_n(0) = u_n(1) = 0, \qquad n = 0, 1, \ldots$$

is a monotone-decreasing sequence converging to the solution $\hat{u}$. Now let

$$S = \{u; u \leq \hat{u} \text{ in } [0, 1]\} \qquad \text{and} \qquad S^* = \{u; u \geq \hat{u} \text{ in } [0, 1]\}$$

For w in S^*, define P_w by $-D^2 + c_w(t, w)$. As we saw before, this is positive-definite. Let us also suppose that $\gamma \geq 0$—that is, $c_u(t, u) \geq 0$—and that $c_{uu}(t, u) \leq 0$. Then for w in S^* and u in S we have

$$\begin{aligned}\langle[f''_{w+\lambda(u-w)} &- P_w](u - w), u - w\rangle \\ &= \langle\{c[t, w + \lambda(u - w)] - c(t, w)\}(u - w), u - w\rangle \geq 0\end{aligned}$$

since $u \leq w$ and $c_{zz} \leq 0$ implies $c[t, w + \lambda(u - w)] \geq c(t, w)$. Thus the hypotheses are satisfied. We now claim that, for w in S^*, the v_w that minimizes $\frac{1}{2}\langle v, P_w v\rangle + \langle\nabla f(w) - P_w w, v\rangle$ over S in fact minimizes it over all H—that is, that the gradient is

$$P_w v + \nabla f(w) - P_w w = 0 \quad \text{at} \quad v = v_w$$

This is well known. To do this, we show that if $P_w v + \nabla f(w) - P_w v = 0$, then v is in S and hence $v = v_w$. This equation for v yields

$$v'' - c_w(t, w)v = c(t, w) - c_w(t, w)w, \qquad v(0) = v(1) = 0$$

which is just the Newton iteration from w to v. Since also

$$\hat{u}'' - c_w(t, w)\hat{u} = c(t, \hat{u}) - c_w(t, w)\hat{u}, \qquad \hat{u}(0) = \hat{u}(1) = 0$$

subtracting we have

$$(v - \hat{u})'' - c_w(t, w)(v - \hat{u}) = c(t, w) - c(t, \hat{u}) - c_w(t, w)(w - \hat{u}) \geq 0$$

since $w \geq \hat{u}$ and $c_{uu} \leq 0$. But then the maximum principle implies that $v - \hat{u} \leq 0$—that is, that v is in S and hence $v = v_w$. Thus we find

$$\begin{aligned}f^*(w) = \tfrac{1}{2}\int_0^1 [v'(t)]^2\, dt &+ \int_0^1 \int_0^{w(t)} c(t, x)\, dx\, dt \\ &+ \tfrac{1}{2}\int_0^1 c_w[t, w(t)][v(t) - w(t)]^2\, dt \\ &+ \int_0^1 c[t, w(t)][v(t) - w(t)]\, dt\end{aligned}$$

where v is the Newton iterate of w solving

$$v'' - c_w(t, w)v = c(t, w) - c_w(t, w)w, \qquad v(0) = v(1) = 0$$

These error bounds are in the norms

$$\int_0^1 \{[e'(t)]^2 + c_z(t, z)e^2(t)\}\, dt \quad \text{for} \quad z = w \quad \text{and} \quad z = \hat{u}$$

Since $c_z(t, z) \geq 0$, we have bounds for $\int_0^1 [e'(t)]^2\, dt$ as well as the fact that $v_w \leq \hat{u} \leq w$. The computable bound for derivatives would thus use the variational results,

$$\tfrac{1}{2}\int_0^1 [\hat{u}'(t) - v_w'(t)]^2\, dt \leq f(v_w) - f(w)$$

General reference: Levitin–Poljak (1966a, b).

2 THEORY OF DISCRETIZATION

2.1. INTRODUCTION

As we remarked in Section 1.1, many problems of practical interest can be considered minimization problems in general spaces. Computationally, however, one is often unable to work in such general spaces since one is restricted usually to dealing with *discrete* data in the real world. While it is useful, as we shall see later, to study methods of minimization in general spaces as well as in finite dimensional spaces, it is also necessary to study the relationships between the solutions of problems in the "discrete" and "continuous" domains. Therefore, we shall now look at this question in a rather general way, to see what kinds of relationships need to exist in order for our computations to be meaningful; later, we shall see how these ideas apply to certain problems.

2.2. CONSTRAINED MINIMIZATION

As usual, we suppose that we seek to minimize the weakly sequentially lower semicontinuous functional f over a weakly sequentially compact subset C of a reflexive space E; an $x^* \in C$ solving this will be called a *solution to the* MPC—the minimization problem over C. We shall hope to compute x^* by dealing with some approximating functionals over approximating spaces, such as quadrature sums for discrete data instead of integrals as in the case of the calculus-of-variations problem.

DEFINITION 2.2.1. A *discretization* for the MPC consists of a family of normed spaces E_n, a family of functionals f_n over E_n, a family of mappings p_n of E_n into E, a family of mapping r_n of E into E_n, and a family of subsets C_n of E_n.

We are thinking of C_n and f_n as "approximations" to C and f, $r_n x \in E_n$ as an "approximation" (restriction) to $x \in E$, and $p_n x_n \in E$ as an "approximation" (prolongation) to $x_n \in E_n$. We shall call x_n^* a *solution to the* MPC$_n$–ϵ_n, the ϵ_n–approximate minimization problem over C_n with $\epsilon_n > 0$ converging to zero, if

$$f_n(x_n^*) \leq \inf_{x_n \in C_n} f_n(x_n) + \epsilon_n$$

For a solution x_n^* to the MPC–ϵ_n to converge in some sense to a solution of the MPC, we need to have some relationships between the two problems to measure the degree of approximation.

DEFINITION 2.2.2. A discretization for the MPC is *consistent* if and only if

1. $\limsup_{n\to\infty} f_n(r_n x^*) \leq f(x^*)$ for some x^* solving the MPC;
2. $\limsup_{n\to\infty} [f(p_n x_n^*) - f_n(x_n^*)] \leq 0$ if x_n^* solves the MPC$_n$–ϵ_n;
3. the sets $C^n \equiv p_n C_n \cup C$ are uniformly bounded, and, if $z_{n_i} \in C^{n_i}$ and $z_{n_i} \rightharpoonup z$, then $z \in C$;
4. solutions $x_{n_i}^*$ of the MPC$_n$–ϵ_n exist for all n;
5. $r_n x^* \in C_n$ for the same x^* in condition 1 above.

We remark that one might of course prove conditions 1 and 2 in Definition 2.2.2 by proving them for all points, not just for the solutions of the minimizing problems. Condition 3 is trivial if $p_n C_n \subset C$. Condition 3 is also trivially true if $e(C^n, C) \to 0$, where $C^n \equiv p_n C_n \cup C$, $e(C^n, C) \equiv \sup_{x \in C^n} d(x, C)$, and $d(x, C) \equiv \inf_{y \in C} \| x - y \|$.

If f is weakly sequentially lower semicontinuous on a set containing, for sufficiently large n, the sets $C^n \equiv p_n C_n \cup C$, then the numbers $\gamma_n \equiv f(x^*) - \inf_{x \in C^n} f(x) \geq 0$ must converge to zero; this follows since, if $f(z_n) \leq \inf_{x \in C^n} f(x) + (1/n)$, there is z in C and a subsequence $z_{n_i} \rightharpoonup z$ yielding $f(x^*) \leq f(z) \leq \liminf_{i\to\infty} f(z_{n_i})$.

EXERCISE. Give a rigorous proof that $\lim_{n\to\infty} \gamma_n = 0$, where γ_n is defined in the preceding paragraph.

We can now prove the following fundamental theorem on approximate minimization via discretizations.

THEOREM 2.2.1. Let f be a weakly sequentially lower semicontinuous functional on a set containing C^n for large n, where $C^n \equiv p_n C_n \cup C$, and let $\{E_n, f_n, C_n, p_n, r_n\}$ be a consistent discretization of the MPC for a weakly sequentially compact set C in a reflexive space E. Let x_n^* and x^* solve the MPC$_n$–ϵ_n and MPC, x^* satisfying conditions 1 and 5 of Definition 2.2.2. Then

$$\lim_{n\to\infty} f(p_n x_n^*) = \lim_{n\to\infty} f_n(x_n^*) = f(x^*)$$

and all weak limit points of $p_n x_n^*$, at least one of which exists, solve the MPC; in particular, if x^* is unique, then $p_n x_n^* \rightharpoonup x^*$.

Proof: Let $\gamma_n \equiv f(x^*) - \inf_{x \in C^n} f(x)$ for large n as above. Since x^* solves the MPC, by the consistency we have

$$f(x^*) \equiv \inf_{x\in C^n} f(x) + \gamma_n \leq f(p_n x_n^*) + \gamma_n \equiv f_n(x_n^*) + \gamma_n + \eta_n$$
$$\eta_n \equiv f(p_n x_n^*) - f_n(x_n^*)$$

where η_n satisfies $\limsup_{n\to\infty} \eta_n \leq 0$ by condition 2 of Definition 2.2.2. On the other hand x_n^* solves the MPC$_n$-ϵ_n, so

$$\begin{aligned} f(x^*) &\leq f_n(x_n^*) + \gamma_n + \eta_n \\ &\leq f_n(r_n x^*) + \gamma_n + \eta_n + \epsilon_n \\ &\equiv f(x^*) + \gamma_n + \eta_n + \epsilon_n + \delta_n \end{aligned}$$

where $\delta_n \equiv f_n(r_n x^*) - f(x^*)$ satisfies $\limsup_{n\to\infty} \delta_n \leq 0$ by condition 1 of Definition 2.2.2. From the first and last terms of this basic inequality we have $\gamma_n + \eta_n + \epsilon_n + \delta_n \geq 0$; since $\lim_{n\to\infty} \epsilon_n = \lim_{n\to\infty} \gamma_n = 0$ while $\limsup_{n\to\infty} \eta_n \leq 0$ and $\limsup_{n\to\infty} \delta_n \leq 0$, we have

$$0 \leq \liminf_{n\to\infty} (\gamma_n + \eta_n + \epsilon_n + \delta_n) \leq \limsup_{n\to\infty} (\gamma_n + \eta_n + \epsilon_n + \delta_n) \leq 0,$$

and therefore $\lim_{n\to\infty} (\eta_n + \delta_n) = 0$. If $\liminf_{n\to\infty} \eta_n \leq \alpha < 0$ for some α, then there exist n_i with $\eta_{n_i} \leq \alpha/2 < 0$ for all i, and hence $\delta_{n_i} \geq -\alpha/4$ for infinitely many i, since $\lim_{n\to\infty} (\eta_n + \delta_n) = 0$, contradicting $\limsup_{n\to\infty} \delta_n \leq 0$. Thus $\lim_{n\to\infty} \eta_n = 0$ and, similarly, $\lim_{n\to\infty} \delta_n = 0$. Letting n now tend to infinity in the basic inequality yields

$$\lim_{n\to\infty} f(p_n x_n^*) = \lim_{n\to\infty} f_n(x_n^*) = f(x^*)$$

Since C^n is uniformly bounded, $\{p_n x_n^*\}$ has weak limit points, all of which lie in C by condition 3 of Definition 2.2.2. For any such limit point z with $p_{n_i} x_{n_i}^* \rightharpoonup z$, $f(z) \leq \liminf_{i\to\infty} f(p_{n_i} x_{n_i}^*) = f(x^*)$, so z must solve the MPC. Q.E.D.

General reference: Daniel (1968b).

2.3. UNCONSTRAINED MINIMIZATION

As we saw in Section 1.4, for the purposes of analysis unconstrained minimization problems are often reduced to constrained problems by means of growth conditions. This same approach is useful for the analysis of constrained minimization via discretization methods. Hence we consider the MPE—the minimization problem over E—to locate x^* with $f(x^*) = \inf_{x \in E} f(x)$ for reflexive E and weakly sequentially lower semicontinuous f. We also consider the discretized problem, the MPE$_n$–ϵ_n, the ϵ_n–approximate minimization problem over E_n, of finding x_n^* in E_n, such that $f_n(x_n^*) \leq \inf_{x_n \in E_n} f_n(x_n) + \epsilon_n$. If x^* exists and the discretization is consistent (here $C_n = E_n$, $C = E$), then the proof of Theorem 2.2.1 with minor modifications shows that $\lim_{n\to\infty} f(p_n x_n^*) = \lim_{n\to\infty} f_n(x_n^*) = f(x^*)$ while, if f satisfies a T-property, one can also conlude that $\{p_n x_n^*\}$ has limit points, all of which solve the MPE.

EXERCISE. State and prove the modification of Theorem 2.2.1 outlined above.

We omit this modification of Theorem 2.2.1, however, because it is generally not useful; the difficulty is that, in practice, one cannot usually prove consistency of a given discretization without having some additional information on the points x_n^*, such as that $\|x_n^*\|_n \leq B$ for a fixed constant B. One way of guaranteeing such a uniform bound is via a uniform-growth condition.

DEFINITION 2.3.1. A discretization for the MPE satisfies a *uniform-growth condition* if and only if

$$\limsup_{n\to\infty} f_n(x_n) = +\infty \quad \text{whenever} \quad \limsup_{n\to\infty} \|x_n\|_n = +\infty$$

Actually, one can describe a uniform T-property, but the above definition covers most of the cases of interest.

DEFINITION 2.3.2. A discretization for the MPE is *stable* if and only if there is a real-valued function $\delta(t)$ for $t \geq 0$, bounded on bounded sets, such that

$$\|x_n\|_n \leq r \quad \text{implies} \quad \|p_n x_n\| \leq \delta(r)$$

Now we can prove the following fundamental theorem on unconstrained minimization via discretization.

THEOREM 2.3.1. Let f be a weakly sequentially lower semicontinuous functional on the reflexive space E, let f satisfy a T-property at 0 with $T = T_0$.

Let the discretization $\{E_n, f_n, p_n, r_n\}$ be stable, be consistent (for uniformly bounded $\|x_n^*\|_n$ and $\|p_n x_n^*\|$), and satisfy a uniform-growth condition. Then

$$\lim_{n\to\infty} f(p_n x_n^*) = \lim_{n\to\infty} f_n(x_n^*) = f(x^*)$$

where x^* solves the MPE, and all weak limit points of $\{p_n x_n^*\}$, at least one of which exists, solve the MPE; in particular, if x^* is unique, then $p_n x_n^* \rightharpoonup x^*$.

Proof: By our assumptions, x_n^* and x^* exist. Since $r_n x^*$ is in E_n,

$$f_n(x_n^*) \leq f_n(r_n x^*) + \epsilon_n \qquad \text{and} \qquad \limsup_{n\to\infty} f_n(r_n x^*) \leq f(x^*)$$

Therefore, there exists a constant $R_1 > 0$ such that

$$f_n(x_n^*) \leq R_1, \qquad f(r_n x^*) \leq R_1 \qquad \text{for all } n$$

Thus, from the uniform-growth condition, there is an $R_2 > 0$ such that

$$\|x_n^*\|_n \leq R_2, \qquad \|r_n x^*\| \leq R_2 \qquad \text{for all } n$$

Hence, by stability,

$$\|p_n x_n^*\| \leq \delta(R_2)$$

Let $R_3 = \max\{T_0, \delta(R_2)\}$, and let

$$C_n = \{x_n; \|x_n\|_n \leq R_2\}$$

and

$$C = \{x; \|x\| \leq R_3\}$$

Then $r_n x^* \in C_n$, $p_n C_n \subset C$, and $x_n^* \in C_n$; therefore, we may restrict ourselves instead to the MPC and MPC$_n$-ϵ_n so that this theorem follows immediately from Theorem 2.2.1. Q.E.D.

Note that the proof of Theorem 2.3.1, above, shows that condition 2 in the consistency definition (Definition 2.2.2) could be proved valid for x_n^* by proving it valid for all sequences $\{x_n\}$ with $\|x_n\|_n$ bounded independently of n.

In some cases one can find estimates for the rate of convergence of $p_n x_n^*$ to x^*. From the proof of Theorem 2.2.1, we see that for large n in the constrained case and all n in the unconstrained case (where $\gamma_n \equiv 0$) we have

$$-\gamma_n \leq f(p_n x_n^*) - f(x^*) \leq \eta_n + \delta_n + \epsilon_n$$

and hence

$$-D_c(n) \leq f(p_n x_n^*) - f(x^*) \leq D_p(n) + D_r(n) + \epsilon_n$$

where D_r, D_c, and D_p measure the defect in consistency via

$$D_r(n) \geq f_n(r_n x^*) - f(x^*)$$
$$D_c(n) \geq f(x^*) - \inf_{C_n} f(x)$$
$$D_p(n) \geq f(p_n x_n^*) - f_n(x^*)$$

In some cases $D_r(n)$, $D_c(n)$, and $D_p(n)$ can be estimated beforehand; if, for example, f involves integration and f_n is a quadrature approximation, $D_r(n)$ and $D_p(n)$ might be estimated from known facts about the accuracy of quadrature formulas. Once we can estimate the speed of convergence of $f(p_n x_n^*)$ to $f(x^*)$, methods such as described by Theorem 1.6.3 can sometimes be used to conclude that $p_n x_n^* \longrightarrow x^*$ and to bound $\| p_n x_n^* - x^* \|$.

The theorems of the previous section, especially Theorem 2.2.1, are related to unpublished work of Aubin and Lions [Aubin–Lions (1966)] treating similar problems. In their work one seeks to minimize $f(x) \equiv J[G(x)]$ over a weakly compact subset C of a reflexive space E, where G maps E into a reflexive space H and is continuous between the weak topologies, and where J is a weakly lower semicontinuous functional on H. Instead, one minimizes $f_n(x_n) = J_n[G_n(x_n)]$ over a weakly compact subset C_n of a reflexive space E^n, where G_n maps E_n into a reflexive space H_n and is continuous between the weak topologies, and where J_n is weakly lower semicontinuous on H_n. Mappings $p_n \colon C_n \longrightarrow C$, $r_n \colon C \longrightarrow C_n$, $q_n \colon H_n \longrightarrow H$, and $s_n \colon H \longrightarrow H_n$ are assumed to exist; the authors make the following assumptions:

1. $|J_n(w_n) - J_n(v_n)| = o(1)$ as $\| w_n - v_n \|_n \longrightarrow 0$;
2. $\lim_{n \to \infty} |J(q_n w_n) - J_n(w_n)| = 0$ if $\| w_n \|_n$ is bounded;
3. $\lim_{n \to \infty} |J_n(s_n w) - J(w)| = 0$ for $w \in H$;
4. If $\| p_n x_n \|$ is bounded, then $G(p_n x_n) - q_n G_n(x_n) \rightharpoonup 0$;
5. $\lim_{n \to \infty} \| G_n(r_n x) - s_n G(x) \|_n = 0$ for $x \in E$;

Under these assumptions the authors show that the x_n^* exactly minimizing f_n over C_n satisfy the properties proved for our x_n^* in Theorem 2.2.1.

EXERCISE. Show that, under hypotheses 1–5, listed above, the discretization $\{E_n, f_n, p_n, r_n, C_n\}$ is consistent and hence Theorem 2.2.1 applies directly to the Aubin–Lions problem.

The special form of f and f_n in the above presentation is related to nonlinear integral equations that are posed as variational problems. We discuss this and other kinds of operator equations briefly.

General reference: Daniel (1968b).

2.4. REMARKS ON OPERATOR EQUATIONS

Suppose one wishes to solve the following nonlinear integral equation [Anselone (1964), Vainberg (1964)]:

$$u(t) = \int_0^1 K(t, \tau)c[\tau, u(\tau)]\, d\tau$$

where we suppose that the integral operator

$$Au \equiv \int_0^1 K(\cdot, \tau)\, u(\tau)\, d\tau$$

is bounded from $L_q(0, 1)$ into $L_p(0, 1)$, where $p \geq 2$, and $1/p + 1/q = 1$; that $K(t, \tau) = K(\tau, t)$, that the spectrum of A as an operator from L_q into L_q is positive and that A maps bounded sets into precompact sets (that is, A is a *compact* operator). We suppose that the operator

$$c(u) \equiv c[\cdot, u(\cdot)]$$

is norm-continuous from $L_p(0, 1)$ into $L_q(0, 1)$—i.e., that $|c(t, u)| \leq a(t) + b|u|^{p/q}$, where $a \in L_q(0, 1)$, and $b > 0$; that $c(t, u)$ is continuous in u for almost all t and measureable in t for all u; and that

$$C(t, u) \equiv \int_0^u c(t, s)\, ds$$

satisfies

$$C(t, u) \leq \alpha u^2 + \beta(t)|u|^\gamma + \delta(t)$$

where

$$\alpha \in (0, m), \qquad m = \inf\{\lambda;\, \lambda \in \sigma(A)\},$$

$$0 \leq \beta \in L_{2/(2-\epsilon)}(0, 1) \quad \text{for some } \epsilon \text{ in } (0, 2), \qquad \text{and} \qquad 0 \leq \delta \in L_1(0, 1)$$

Then we can write

$$A = GG^*$$

where G is positive and compact from $L_2(0, 1)$ into $L_p(0, 1)$, and deduce that the functional

$$f(x) = \langle x, x\rangle - 2\int_0^1 C[t, Gx(t)]\, dt$$

is defined on $L_2(0, 1)$ and achieves its minimum there. At the minimum x^*, ∇f vanishes, and so

$$0 = \nabla f(x^*) = 2x^* - 2G^* c(Gx^*)$$

Defining $u^* = Gx^* \in L_p(0, 1)$, we see that

$$u^* = Ac(u^*)$$

and that u^* solves the integral equation.

If we define

$$J(w) = \langle G^{-1}w, G^{-1}w\rangle - 2\int_0^1 C[t, w(t)]\,dt$$

for $w \in L_p(0, 1)$, we see that

$$f(x) = J[G(x)]$$

as described by the Aubin–Lions work. Thus the theory described in Section 2.3 applies to numerical solution of integral equations, where integration is, for example, discretized by means of quadrature formulas.

A particularly attractive feature of integral equations is the compactness of the operator A. If A is approximated by a quadrature sum,

$$A_n u = \sum_{i=1}^{n} w_{n,i} K(\cdot, \tau_i) u(\tau_i) \equiv Q_n[Ku]$$

where the quadrature formula Q_n satisfies

$$\lim_{n \to \infty} Q_n[f] = \int_0^1 f(t)\,dt$$

for all continuous f, then the operators A_n turn out to be *collectively compact* in many cases; that is, the union of the images by A_n of each bounded set is compact. This fact has been exploited greatly [Anselone (1965, 1967), Anselone–Moore (1964), Moore (1966)] to analyze numerical methods for linear integral equations. Essentially, the same viewpoint has been used to analyze nonlinear equations given by variational problems [Daniel (1968a)]. A typical result using this viewpoint is as follows.

If f and f_n, $n = 1, 2, \ldots$, are weakly lower semicontinuous functionals such that for each x in a weakly compact subset C of a reflexive space E we have

$$\lim_{n \to \infty} f_n(x) = f(x)$$

and such that $\{\nabla f - \nabla f_n\}$ is a collectively compact set of norm-continuous mappings of E into E^*, then if $x_n^* \in C$ satisfies $f_n(x_n^*) \leq \inf_{x \in C} f_n(x) + \epsilon_n$, with $\epsilon_n \longrightarrow 0$, it follows that every weak limit point of $\{x_n^*\}$, at least one of which exists, minimizes f over C.

EXERCISE. Show that the collective compactness referred to above merely serves to guarantee the consistency of the discretization with $C_n = C$, $E_n = E$, $p_n = r_n =$ the identity map.

Thus most results of Daniel (1968a) follow from either of the above Theorems 2.2.1 or 2.3.1.

It is a trivial exercise further to deduce from these theorems results concerning the solution of operator equations via discretizations; one need only recall that $\nabla f(x^*) = 0$ at an interior minimum of f. Thus one is led to results concerning the *weak* convergence of $p_n x_n^*$ to x^* where

$$\nabla f_n(x_n^*) = 0 \qquad \text{and} \qquad \nabla f(x^*) = 0$$

Stronger convexity hypotheses on f will then give norm convergence.

The analysis of convergence for discretization methods of solution of nonlinear equations has been carried much further, however, than can be covered from the variational viewpoint. Rather than give such an incomplete picture of the subject, therefore, we merely refer the interested reader to the literature. We proceed, in the following chapter, to examine a number of examples to which the variational viewpoint naturally applies in order to demonstrate some particular cases of discretization methods.

General references: Aubin (1967a, b; 1968), Browder (1967), Petryshyn (1968).

3 EXAMPLES OF DISCRETIZATION

3.1. INTRODUCTION

In the well-developed theory of discretizations for operator equations, many examples of particular discretization schemes can be found, particularly for partial differential equations (see the General References at the end of this section for such general examples). In this chapter we shall examine some specific types of problems or methods which, by our considering a particular form for the discretization, can be analyzed from the viewpoint of the discretization theory of variational problems and from the theorems presented in Chapter 2 or extensions of those theorems. In some cases this leads to new results, in some it provides a different way of looking at well-known results, and in one it shows how the approach can be used to guide the direction of one's research on a new method.

General references: Aubin (1967a, b; 1968).

3.2. REGULARIZATION

The idea of *regularization* has been studied from at least two different viewpoints. Under the name of *regularization* it was developed theoretically largely by the Russian school [Levitin–Poljak (1966b), Tikhonov (1965)] for the situation in which one seeks to minimize a functional g and, out of all the solutions to this problem, find the one which is "smoothest" or "most regular" with respect to another functional h—that is, which minimizes h over the set of solutions of the first problem. If g represents a calculus-of-variations problem, for example, one might take

$$h(x) = \int_0^1 |\dot{x}(t)|^2 \, dt$$

to make x "smooth." This goal can be accomplished in many cases by minimizing

$$g + a_n h, \qquad \text{where} \qquad a_n > 0, \quad \text{and} \lim_{n\to\infty} a_n = 0 \tag{3.2.1}$$

and noting that the solutions to these problems converge to the desired regular solution. This same technique has also been studied as a form of the *penalty-function method*, since minimization of the functional in Equation 3.2.1 is equivalent to minimizing

$$h + \frac{1}{a_n} g, \qquad \text{where} \qquad a_n > 0, \qquad \text{and} \qquad \lim_{n\to\infty} a_n = 0$$

In this form we recognize the procedure as a form of the penalty-function technique to minimize h over the set of x satisfying $g(x) = 0$, if $g(x) \geq 0$ for all x [Courant (1943), Butler–Martin (1962)].

We shall briefly consider this method (from the regularization viewpoint) as a discretization. First we shall generalize it somewhat because of its relevance for numerical work, and then we shall specialize to the above description.

Suppose we seek to minimize the nonnegative weakly sequentially lower semicontinuous functional h over the set of points which minimize the weakly sequentially lower semicontinuous functional g over a weakly sequentially compact subset C of a reflexive space E. Suppose we have a discretization for this problem described by $\{E_n, g_n, h_n, C_n, p_n, r_n\}$. For a sequence of positive a_n tending to zero we shall define

$$f_n(x_n) = g_n(x_n) + a_n h_n(x_n), \qquad \text{for} \qquad x_n \in E_n$$

We also define

$$f(x) = g(x), \qquad \text{for} \qquad x \in E$$

and thus we have a discretization $\{E_n, f_n, C_n, p_n, r_n\}$. We list assumptions for this example corresponding to the consistency definition (Definition 2.2.2), but stronger:

1. $\limsup_{n\to\infty} [g_n(r_n x^*) + a_n h_n(r_n x^*) = g(x^*) - a_n h(x^*)] \equiv \limsup_{n\to\infty} \zeta_n \geq 0$ for every x^* minimizing g over C;
2. $\limsup_{n\to\infty} [g(p_n x_n^*) + a_n h(p_n x_n^*) - g_n(x_n^*) - a_n h_n(x_n^*)] \equiv \limsup_{n\to\infty} \beta_n \leq 0$ if x_n^* ϵ_n–approximately minimizes f_n over C_n;
3. the sets $C^n \equiv p_n C_n \cup C$ are uniformly bounded, and, if $z_{n_i} \in C^{n_i}$ with $z_{n_i} \rightharpoonup z$, then $z \in C$;

4. solutions x_n^* exist;
5. $r_n x^* \in C_n$ for every x^* minimizing g over C.

THEOREM 3.2.1. Suppose x_n^* satisfies

$$g_n(x_n^*) + a_n h_n(x_n^*) \leq \inf_{x_n \in C_n} [g_n(x_n) + a_n h_n(x_n)] + \epsilon_n$$

Under hypotheses 1–5 (above) on the discretization for the weakly sequentially lower semicontinuous functionals g and h, with $h \geq 0$, $h_n \geq 0$, and C weakly sequentially compact, if in addition $a_n > 0$ converges to zero slowly enough that

$$\limsup_{n \to \infty} \frac{\zeta_n + \beta_n + \epsilon_n + \gamma_n}{a_n} \leq 0$$

where $\gamma_n \equiv g(x^*) - \inf_{C_n} g(x)$, then all weak limit points x' of $p_n x_n^*$, at least one of which exists, minimize h over the set of minimizing points of g over C.

Proof: Since $h_n \geq 0$ and $h \geq 0$, it is trivial to verify that hypotheses 1–5 (above) imply the consistency of the discretization $\{E_n, f_n, C_n, p_n, r_n\}$ for minimizing $f = g$ over C.

EXERCISE. Prove that hypotheses 1–5 (above) imply the consistency of the discretization $\{E_n, f_n, C_n, p_n, r_n\}$ for minimizing $f = g$ over C.

Thus by Theorem 2.2.1, weak limits x' exist and all minimize $f = g$ over C. As in the proof of Theorem 2.2.1, for large n we have $g(x^*) \leq g(p_n x_n^*) + \gamma_n$ for any x^* minimizing g over C. Thus for large n, we write

$$\begin{aligned}
&g(x^*) \leq g(x^*) + a_n h(p_n x_n^*) \leq g(p_n x_n^*) + a_n h(p_n x_n^*) + \gamma_n \\
&= g_n(x_n^*) + a_n h_n(x_n^*) + \gamma_n + \beta_n \leq g_n(r_n x^*) + a_n h_n(r_n x^*) + \gamma_n + \epsilon_n + \beta_n \\
&= g(x^*) + a_n h(x^*) + \gamma_n + \epsilon_n + \beta_n + \zeta_n
\end{aligned}$$

Therefore,

$$g(x^*) + a_n h(p_n x_n^*) \leq g(x^*) + a_n h(x^*) + \gamma_n + \epsilon_n + \beta_n + \zeta_n$$

which implies

$$h(p_n x_n^*) \leq h(x^*) + \frac{\epsilon_n + \beta_n + \zeta_n + \gamma_n}{a_n}$$

Thus, if $p_{n_i} x_{n_i}^* \rightharpoonup x'$, we have

$$h(x') \leq \liminf_{i \to \infty} h(p_{n_i} x_{n_i}) \leq h(x^*)$$

Thus x' minimizes h over the set of minimizing points of g. Q.E.D.

COROLLARY 3.2.1. If g and h are weakly sequentially lower semicontinuous functionals over the weakly sequentially compact subset C of a reflexive space E, with $h \geq 0$, and if x_n^* satisfies

$$g(x_n^*) + a_n h(x_n^*) \leq \inf_{x \in C} [g(x) + a_n h(x)] + a_n \delta_n$$

where $a_n, \delta_n > 0$, $\lim_{n\to\infty} a_n = \lim_{n\to\infty} \delta_n = 0$, then all weak limit points x' of x_n^*, at least one of which exists, minimize h over the set of minimizing points of g over C.

Proof: Let $E_n = E$, $C_n = C$, $p_n = r_n =$ the identity map, $g_n = g$, and $h_n = h$; the hypotheses in and immediately preceding Theorem 3.2.1 are clearly satisfied, since $g + a_n h$ is weakly sequentially lower semicontinuous. Q.E.D.

The above corollary describes the nature of the regularization method as it is most often described [Levitin–Poljak (1966b)].

It is possible in many cases to guarantee more than the rather weak convergence properties guaranteed in Theorem 3.2.1; we give an example below.

THEOREM 3.2.2. If, in addition to the hypotheses in and immediately preceding Theorem 3.2.1, C is convex, g is quasi-convex, and either g or h is strongly quasi-convex, then the entire sequence $\{p_n x_n^*\}$ is weakly convergent. If, in addition, either g or h is uniformly quasi-convex, the sequence is norm-convergent.

Proof: If g is strongly quasi-convex, then by Theorem 1.5.2 the set

$$C' = \{x^*; x^* \in C, g(x^*) = \inf_{x \in C} g(x)\}$$

consists of only one point, so $p_n x_n^* \rightharpoonup x^*$. If g is only quasi-convex, then C' is convex, and by Theorem 1.5.2 the strongly quasi-convex functional h is minimized over C' at a unique point x', so $p_n x_n^* \rightharpoonup x'$. If g is uniformly quasi-convex in addition, then $p_n x_n^* \rightarrow x^*$ by Theorem 1.6.1. If only h is uniformly quasi-convex, then we write

$$\delta(\| x' - p_n x_n^* \|) \leq \max \{h(x'), h(p_n x_n^*)\} - h\left(\frac{x' + p_n x_n^*}{2}\right)$$

Recalling from the proof of Theorem 3.2.1 that

$$h(p_n x_n^*) \leq h(x') + \frac{\epsilon_n + \beta_n + \zeta_n + \gamma_n}{a_n}$$

we have

$$\limsup_{n\to\infty} \delta(\|x' - p_n x_n^*\|) \leq \limsup_{n\to\infty} \left\{\max\left[h(x'), h(x') + \frac{\epsilon_n + \beta_n + \zeta_n + \gamma_n}{a_n}\right] - h\left(\frac{x' + p_n x_n^*}{2}\right)\right\} = h(x') - \liminf_{n\to\infty} h\left(\frac{x' + p_n x_n^*}{2}\right) \leq 0$$

since $(x' + p_n x_n^*)/2 \rightharpoonup x'$, and hence $p_n x_n^* \rightarrow x'$. Q.E.D.

In some cases one can also compute the order of accuracy of the regularized solution as a function of the parameter a_n; in the following paragraph we briefly describe some recent results of this type [Aubin (1969a, 1969b)] which have application to optimal-control problems.

Suppose we wish to minimize the convex and differentiable functional f over the set $C \equiv \{x; Lx = b\}$ where b is given and L is a bounded linear operator from the Hilbert space E_1 into a Hilbert space E_2 such that L has a closed range. To do this we instead minimize $f_n(x) \equiv \|Lx - b\|^2 + a_n f(x)$ over E_1. Suppose that x^* minimizes f over C and x_n^* minimizes $\|Lx - b\|^2 + a_n f(x)$ over all of E_1. Then it can be proved [Aubin (1969a)] that the following estimates hold for some constant $k > 0$: (1) $\|b - Lx_n\| \leq ka_n$; and (2) $f(x^*) - (1/a_n)f_n(x_n^*) \leq f(x^*) - f(x_n^*) \leq ka_n$. Under stronger hypotheses on f, this of course leads to error bounds for $\|x_n^* - x^*\|$. For computational purposes, if the problem is discretized via mappings p_n, r_n—for example, by using finite differences to replace the differential equations $Lx = b$ of a control-theory problem—the same type of error estimates are known as those given above [Aubin (1969b)].

3.3. A NUMERICAL METHOD FOR OPTIMAL-CONTROL PROBLEMS

We seek to compute numerically an approximate solution to an optimal-control (C-problem) of the following type: minimize

$$f(y, u) = \int_{t_0}^{t_1} c[t, y(t), u(t)]\, dt$$

where the cost function $c(t, y, u)$ is nonnegative, over the collection of functions (y, u) satisfying

$$\dot{y} \equiv \frac{dy}{dt} = s[t, y(t), u(t)], \qquad y(t) \in Y(t), \qquad u(t) \in U(t),$$

$$t_0 \leq t \leq t_1, \qquad y(t_0) \in Y_I \quad \text{and} \quad y(t_1) \in Y_F$$

where t_0 and t_1 are unknown points in some fixed interval $[0, T]$ and Y_I, Y_F, $Y(t)$, and $U(t)$ are specified subsets of E^l, E^l, E^l, and E^k, respectively. As is well known [Warga (1962)], this problem can be transformed to one with *fixed time*—i.e., to $t_0 = 0$, $t_1 = 1$—essentially by introducing t_0 and t_1 as components of an "extended" y-vector; this transformation preserves important properties of the problem, including the form of the y- and u-constraints, so hereafter we shall assume that we have a fixed-time problem with $t_0 = 0$, $t_1 = 1$.

ASSUMPTION A1. $t_0 = 0$, $t_1 = 1$.

EXERCISE. Supply the details to justify the above specialization to $t_0 = 0$, $t_1 = 1$.

The following numerical method has been proposed to solve the C-problem [Rosen (1966)]: for positive integers n, set $k \equiv k_n = 1/n$, $t_i = ik$ for $0 \leq i \leq n$; find vectors $y_n = (y_{n,0}, \ldots, y_{n,n})$, $u_n = (u_{n,0}, \ldots, u_{n,n})$ minimizing $k \sum_{i=1}^{n} c(t_i, y_{n,i}, u_{n,i})$ over the collection of vectors satisfying

$$\frac{y_{n,i+1} - y_{n,i}}{k} = s(t_i, y_{n,i}, u_{n,i}) \quad \text{for} \quad i = 0, \ldots, n-1, \quad y_{n,i} \in Y(t_i)$$

$$\text{and } u_{n,i} \in U(t_i) \quad \text{for} \quad i = 0, \ldots, n, y_{n,0} \in Y_I, y_{n,n} \in Y_F$$

This method has proved useful in practice; under certain assumptions [Rosen (1966)], the nonlinear *programming problem* (*P-problem*) defined by the numerical approximation can be computed rapidly by a variant of Newton's method. We are concerned not with methods for computing (y_n, u_n) but with whether or not the sequence (y_n, u_n) or—more precisely—approximations to (y_n, u_n) converge in some sense to a solution (y, u) to the original C-problem. In Cullum (1969), some results are obtained concerning this convergence, particularly for C-problems with $s(t, y, u)$ linear in y and u; in a certain sense which will become clear later, the convergence statements of that approach do not quite face the computational problems squarely (except for problems lacking state constraints [Cullum(1970)]). We shall examine the method of Rosen (1966) in detail and see that the convergence theory can be treated nicely by the discretization approach. In fact, Rosen (1966) treats the problem computationally as one involving *inequality* constraints

$$y_{n,i+1} \leq y_{n,i} + ks(t_i, y_{n,i}, u_{n,i})$$

and indicates that one can solve the problem under *equality* constraints by a penalty-function approach. This allows us to analyze the method by means of the tools of regularization as discussed in Section 3.2 and, in particular, in Theorem 3.2.1. Therefore, for a sequence of positive numbers a_n converging to zero, we shall approximately minimize

$$k \sum_{i=1}^{n} \left[s(t_{i-1}, y_{n,i-1}, u_{n,i-1}) - \frac{y_{n,i} - y_{n,i-1}}{k} \right] + a_n k \sum_{i=1}^{n} c(t_i, y_{n,i}, u_{n,i})$$

under the above *inequality* constraints and with the sets $Y(t_i)$, $U(t_i)$ slightly expanded; the points obtained will be shown to converge to a solution of the control problem under *equality* constraints. The sets $Y(t_i)$ and $U(t_i)$ must be expanded slightly in order to guarantee in general the existence of feasible points for the discrete P-problem arbitrarily near the solution to the C-problem. To see the reason for this, consider the following C-problem: Solve $\dot{y} = u$, $t^2 \leq y \leq t^2 + t$, $2t \leq u \leq 3t$, $t \in [0, 1]$, minimizing $\int_0^1 (y^2 + u^2)\, dt$. The solution is $y = t^2$, $u = 2t$; but there are no feasible points at all for the P-problem with constraints $y_{i+1} = y_i + ku_i$, $(ik)^2 \leq y_i \leq (ik)^2 + ik$, $2ik \leq u_i \leq 3ik$, because $y_0 = u_0 = 0$ implies $y_1 = 0$. In this example, however, it is clear that there exist points satisfying the equality constraints for the P-problem and which are very near to being in $Y(t_i)$, $U(t_i)$.

EXERCISE. For the example immediately above, show that there exist points satisfying the equality constraints which are very "near" to being in $Y(t_i)$ and $U(t_i)$ for all i, where "near" is some reasonable and precise concept.

The assumption in general that this is true is essentially equivalent to the assumption of the existence of a mapping r_n in a consistent discretization scheme.

Continuing with the intuitive approach, we also see that in order for the numerical method to have a chance of success, the nature of the sets $Y(t)$, $U(t)$ must be revealed fully by their nature at the discrete points t_i; for example, if $Y(t) = \{t\}$ for irrational t, but $Y(t) = (-\infty, \infty)$ for rational t, then the numerical method using $Y(i/n)$ would never detect restrictions. Thus we need to assume that $Y(t)$, $U(t)$ vary nicely in the sense that, given feasible vectors for the P-problem (y_n, u_n) with $y_{n,i} \in Y(t_i)$, $u_{n,i} \in U(t_i)$, then there exist feasible functions (y, u) for the C-problem with $y(t)$ near $Y(t)$, $u(t)$ near $U(t)$, and y and u near y_n and u_n in some sense; the point (y, u) will be called $p_n(y_n, u_n)$, where p_n will turn out to be the relevant mapping in a consistent discretization.

For notational convenience, we restrict ourselves henceforth to scalar problems; that is, we assume that y and u are in E^1. The situation for $y \in E^l$, $u \in E^k$ is exactly the same except that some statements—such as those regarding convexity of functions $s(t, y, u)$—must be read with regard to the vector-valued function's individual components.

Now we are ready to make our intuitive assumptions more precise. Define the Hilbert space $E = \{(y, u); u \in L_2(0, 1), \dot{y} \in L_2(0, 1), y \text{ is absolutely continuous}\}$. For $x = (y, u) \in E$, let $\|x\|^2 = \int_0^1 \dot{y}^2\, dt + \int_0^1 u^2\, dt + y(0)^2$. This is essentially a standard Sobolev space. Define the discretized space

$$E_n = \{(y_n, u_n);\ y_n = (y_{n,0}, \ldots, y_{n,n}),\ u_n = (u_{n,0}, \ldots, u_{n,n})\}$$

For $x_n = (y_n, u_n) \in E_n$, let

$$\|x_n\|_n^2 = k \sum_{i=1}^{n} \left(\frac{y_{n,i} - y_{n,i-1}}{k}\right)^2 + k \sum_{i=1}^{n} u_{n,i}^2 + y_{n,0}^2$$

As we noted in Section 1.3, weak convergence in E is equivalent to weak convergence of the components $\dot{y}$ and u in $L_2(0, 1)$ and convergence of $y(0)$ in $\mathbb{R}$, which implies uniform convergence of y—i.e., convergence in $C[0, 1]$.

Next we define functionals $h(x)$, $g(x)$, $h_n(x_n)$, $g_n(x_n)$ for $x = (y, u)$, $x_n = (y_n, u_n)$ as follows:

$$h(x) = \int_0^1 c[t, y(t), u(t)]\, dt$$

$$g(x) = \int_0^1 \{s[t, y(t), u(t)] - \dot{y}(t)\}\, dt$$

$$h_n(x_n) = k \sum_{i=1}^{n} c(t_i, y_{n,i}, u_{n,i})$$

$$g_n(x_n) = k \sum_{i=1}^{n} \left[s(t_{i-1}, y_{n,i-1}, u_{n,i-1}) - \frac{y_{n,i} - y_{n,i-1}}{k} \right]$$

Let

$$Q' = \{(y, u) \in E; y(t) \in Y(t) \text{ for all } t \in [0, 1], y(0) \in Y_I, y(1) \in Y_F\}$$

$$Q'' = \{(y, u) \in E; u(t) \in U(t), \dot{y}(t) \le s[t, y(t), u(t)] \text{ for almost all } t \in [0, 1]\}$$

$$Q''' = \{(y, u) \in E; g(y, u) = 0\}$$

Our C-problem now takes the following form: find an $x^* = (y^*, u^*) \in Q_0 = Q' \cap Q'' \cap Q'''$ satisfying $h(x^*) = h(y^*, u^*) = \inf_{x \in Q_0} h(x)$.

ASSUMPTION A2. We assume there exists $\delta_0 > 0$ such that for all $\delta \in [0, \delta_0)$, the set $Q(\delta) \equiv \{(y, u) \in E; d[y(t), Y(t)] \le \delta$ for all $t \in [0, 1]$, $d[y(0), Y_I] \le \delta$, $d[y(1), Y_F] \le \delta$, $d[u(t), U(t)] \le \delta$ and $\dot{y}(t) \le s[t, y(t), u(t)]$ for almost all $t \in [0, 1]\}$ is weakly sequentially compact and bounded by the constant B.

The boundedness of $Q(\delta)$ can be deduced if, for example, the sets $Y(t)$, $U(t)$ are bounded above and below by functions in $L_2(0, 1)$ or if in some other fashion one can find a priori bounds on the solutions and then include the bounds (theoretically) in the constraints. If, furthermore, the set of (y, u) satisfying the differential inequalities forms a weakly closed subset and if Y_I, Y_F, $Y(t)$, and $U(t)$ are closed for each t and $U(t)$ is a convex set for each t, it is easy to deduce that the set of (y, u) satisfying the other constraints is weakly closed and hence $Q(\delta)$ is a weakly closed subset of a bounded set and is therefore weakly sequentially compact.

EXERCISE. Supply the details for the above argument concerning the weak sequential compactness of $Q(\delta)$.

ASSUMPTION A3. Suppose h and g are weakly sequentially lower semicontinuous functionals and that x^* solves the C-problem—i.e., $x^* \in Q_0$, $h(x^*) = \inf_{x \in Q_0} h(x)$.

ASSUMPTION A4. Assume there exists a map r_n of E into E_n such that, for *some* x^* solving the C-problem, $(y_n, u_n) \equiv x_n \equiv r_n x^*$ satisfies the following:

1. $\lim_{n\to\infty} h_n(x_n) = h(x^*)$;
2. $y_{n,i+1} = y_{n,i} + ks(t_i, y_{n,i}, u_{n,i})$ for $0 \le i \le n-1$, $y_{n,0} \in Y_I$;
3. $\lim_{n\to\infty} d_n = 0$, where $d_n \ge d(x_n) \equiv \max_{0\le i\le n} \max \{d[y_{n,i}, Y(t_i)], d(y_{n,n}, Y_F), d[u_{n,i}, U(t_i)]\}$.

We define the expanded constraint sets for the P-problem now as

$$Q_n = \{x_n = (y_n, u_n) \in E_n; ||x_n||_n \le B + d_n, d(x_n) \le d_n, \\ y_{n,i+1} \le y_{n,i} + ks(t_i, y_{n,i}, u_{u,i}) \text{ for } 0 \le i \le n-1, y_{n,0} \in Y_I\}$$

Our P-problem will be to approximately minimize $g_n(x_n) + a_n h_n(x_n)$ over Q_n.

ASSUMPTION A5. Assume there exists a map p_n of E_n into E such that, if $x_n \in Q_n$, then

1. $\lim_{n\to\infty} |h_n(x_n) - h(p_n x_n)| = \lim_{n\to\infty} |g_n(x_n) - g(p_n x_n)| = 0$;
2. $(z_n, w_n) \equiv p_n x_n$ satisfies $\dot{z}_n(t) \le s[t, z_n(t), w_n(t)]$, for almost all $t \in [0, 1]$, $z_n(0) \in Y_I$;
3. $\lim_{n\to\infty} e_n = 0$, where $e_n \ge e(p_n x_n) \equiv \sup_{0\le t\le 1} \max \{d[z_n(t), Y(t)], d[z_n(1), Y_F], d[w_n(t), U(t)]\}$, and $e_{n+1} \le e_n$;
4. $||p_n x_n|| \le B + e_n$.

Finally, we define the slightly enlarged constraint set Q^n for the C-problem:

$$Q^n \equiv \{x = (y, u) \in E; ||x|| \le B + e_n, e(x) \le e_n, \dot{y}(t) \le s[t, y(t), u(t)] \\ \text{for almost all } t \in [0, 1], y(0) \in Y_I\}.$$

Under all the above sets of assumptions, we can now prove that sufficiently accurate approximate solutions to the penalty-function form of the P-problem will converge to a solution of the C-problem, by using Theorem 3.2.1; we shall later examine hypotheses under which Assumptions A1–A5 will be valid.

THEOREM 3.3.1. Let Assumptions A1–A5 hold, let $h, g, h_n, g_n, Q_0, Q_n, Q^n, r_n, p_n$ be as described above; and let $a_n > 0$, $\lim_{n\to\infty} a_n = 0$. For each n, let x_n^* satisfy

$$g_n(x_n^*) + a_n h_n(x_n^*) \leq g_n(x_n) + a_n h_n(x_n) + a_n \delta_n \qquad \text{for all } x_n \in Q_n$$

where $\delta_n \geq 0$, $\lim_{n\to\infty} \delta_n = 0$. Then all weak limit points x' of $p_n x_n^*$, at least one of which exists, solve the C-problem—i.e., if $x' = (y, u)$, then $\dot{y} = s[t, y(t), u(t)]$ almost everywhere, $x' \in Q_0$, and $h(x') \leq h(x)$ for all x in Q_0.

Proof: We wish to apply Theorem 3.2.1, if possible; we check five numbered hypotheses preceding that theorem with $C \equiv Q_0, C_n \equiv Q_n$. Number 1 is true by Assumption A4, but only for *some* x^*, not *all* x^*; No. 2 is valid by Assumption A5; No. 4 is assumed above; No.5 is valid by Assumption A4, but only for *some*, not *all*, x^*. For condition 3, we note that $C^n \equiv p_n C_n \cup C \subset Q^n$. If $z_{n_i} \in C^{n_i}$ and $z_{n_i} \rightharpoonup z$, since $Q^{n+1} \subset Q^n$ and Q^n is weakly sequentially compact because of Assumption A2, we conclude that $z \in Q^{n_i}$ for all i and hence $z \in \bigcap_1^\infty Q^n \subset Q_0$, as demanded by condition 3. Although we cannot exactly apply Theorem 3.2.1, we can follow the lines of its proof, making use of additional information we have in this case.

Thus we can conclude, as in Theorem 3.2.1, that a weak limit point x' of $\{p_n x_n\}$ exists, must lie in Q_0, and minimizes g over C—that is, $g(x') = 0$. The hypotheses 1 through 5 and that on the decay rate of a_n were used in Theorem 3.2.1 only to show that $h(x') \leq h(x^*)$ (where x^* solves the C-problem in our case); we can handle this differently. We have

$$\begin{aligned} 0 = g_n(r_n x^*) \leq g_n(x_n^*) \leq g_n(x_n^*) + a_n h_n(x_n^*) &\leq g_n(r_n x^*) + a_n h_n(r_n x_n^*) \\ &+ a_n \delta_n \leq g_n(x_n^*) + a_n h_n(r_n x^*) + a_n \delta_n \end{aligned}$$

which implies

$$h_n(x_n^*) \leq h_n(r_n x^*) + \delta_n$$

Then

$$\begin{aligned} h(x') &\leq \liminf_{n\to\infty} h(p_n x_n^*) = \liminf_{n\to\infty} h_n(x_n^*) \\ &\leq \liminf_{n\to\infty} [h_n(r_n x^*) + \delta_n] = h(x^*) \end{aligned}$$

Since $h(x') \leq h(x^*)$ and x^* minimizes h, so must x'. Q.E.D.

EXERCISE. Provide all the details for the *Proof* of Theorem 3.3.1 above.

The results of Section 3.2 concerning stronger convergence properties of course apply here also, but we shall not state them again. Rather we must consider conditions under which the assumptions in the theorem are true. The theorem above (Theorem 3.3.1) merely serves to identify conditions sufficient to guarantee the applicability of the numerical method of Rosen (1966). We note that the existence of p_n is important only to the proof, while the existence of r_n and the numbers d_n related thereto are crucial to the numerical algorithm itself; we are required to treat the P-problem over Q_n, a set defined via d_n, and we must therefore know d_n in order actually to compute. In Cullum (1969), it is shown that, for certain problems, if the sets $Y(t)$ are expanded by distances γ_n, $U(t)$ by distances σ_l, and a discretized step size k of length $k = 1/m$ is used, then sequences $m(n)$ and $l(n)$ exist such that maps p_n, r_n exist for the problem defined by γ_n, $\sigma_{l(n)}$, $k_n = 1/m(n)$, with $d_n \leq \gamma_n + \sigma_{l(n)}$. This does not really yield a computational procedure, since for a given sequence of step sizes k we still do not know by how much to expand the constraint sets.

Now we shall attempt to make the numerical method really implementable; another approach to this can be found in Cullum (1970) for problems lacking state constraints. First, however, we remark that the assumptions other than A4 and A5 are reasonable assumptions insofar as the existence of the solution to the C-problem and the computability of approximate solutions of the P-problem are concerned. In Rosen (1966), for example, in order to prove that the numerical method used there for minimizing $g_n + a_n h_n$ works, it is assumed that $s(t, y, u)$ and $c(t, y, u)$ are convex jointly in y and u; it is a simple matter to show—using this assumption, the assumptions in the paragraph following A2, and the additional one that $s_y(t, y, u)$, $s_u(t, y, u)$, $c_y(t, y, u)$, $c_u(t, y, u)$ exist and as functions of t are in $L_2(0, 1)$ for fixed $(y, u) \in E$—that f, h, Q satisfy their needed assumptions. We therefore do not discuss these assumptions further.

EXERCISE. Indicate how the assumptions of the preceding paragraph can be used to deduce that f, h, and Q satisfy the assumptions demanded by the theory developed so far.

Let us consider the mapping p_n; we must apply it to points $x_n = (y_n, u_n)$ satisfying

$$y_{n,i+1} = y_{n,i} + ks(t_i, y_{n,i}, u_{n,i}) - kb_{n,i} \text{ with } b_{n,i} \geq 0 \text{ for } 0 \leq i \leq n-1$$

If we define $w_n(t) = p_n u_n$ as a step function constant on each interval (t_i, t_{i+1}) with value $u_{n,i}$ and $b_n(t)$ similarly, then y_n looks like the numerical solution of the equation $\dot{z} = s[t, z(t), w_n(t)] - b_n(t)$, $z(0) = y_{n,0}$; if we define $v_n(t) = p_n y_n$ as the solution of this equation, then we are asking y_n and $v_n(t)$ to be close in some sense uniformly in u_n and b_n. Even then, one needs to know that $Y(t)$ and $U(t)$ are continuous enough that $v_n(t_i)$ near $Y(t_i)$ for all i will imply the nearness for all t, and similarly for w_n, U. Finally, to conclude satisfying Assumption A5, we need $h(p_nV_n) - h_n(V_n)$ and $g(p_nV_n) - g_n(V_n)$ to tend to zero. We give some conditions under which Assumption A5 is valid via this approach.

For any set T and positive number ϵ, let $N(T, \epsilon) = \{z; d(z, T) < \epsilon\}$. We shall say a *set function* $T(t)$ is *continuous* on $0 \leq t \leq 1$ if and only if for $\epsilon > 0$ there exists $\delta > 0$ such that $|t' - t''| < \delta$ implies $T(t') \subset N[T(t''), \epsilon]$.

ASSUMPTION A6. Assume that $Y(t)$ and $U(t)$ are continuous set functions.

ASSUMPTION A7. Suppose that for each $w \in L_2(0, 1)$ with $w(t) \in U(t)$ almost everywhere and $z(0) \in Y_I$, there exists a unique solution of $\dot{z}(t) = s[t, z(t), w(t)]$, $z(0) = 0$ for almost all $t \in [0, 1]$, and that the set of such solutions $z(t)$ is bounded uniformly in such w and z_0.

ASSUMPTION A8. Assume there exists a function $q(t, y)$ continuous in (t, y) for (t, y) in $[0, 1] \times (-\infty, \infty)$ and such that, if $l(t, y, u)$ is either of the functions $s(t, y, u)$ or $c(t, y, u)$, we have

$$|l(t', y', u) - l(t'', y'', u)| \leq |q(t', y') - q(t'', y'')|$$

for all $u \in U^* = \{u; u \in U(t) \text{ for some } t \in [0, 1]\}$.

Remark: If both $Y(t)$ and $U(t)$ are of the form $Y(t) = \{y; m(t) \leq y \leq M(t)\}$ for continuous m, M, then they are continuous set functions. If U^* and Y are compact, if $s(t, y, u)$ is Lipschitz-continuous in y uniformly in $(t, u) \in [0, 1] \times U^*$, and if $|s(t, y, u)| \leq \mu(t)\,\alpha(|y|)$ for $u \in U^*$ where $\mu(t)$ is integrable on $[0, 1]$ and $\alpha(|y|) = O(|y|)$ as $|y| \longrightarrow \infty$, then Assumption A7 is valid [Roxin (1962)].

EXERCISE. Prove that a set function of the form $Y(t) = \{y; m(t) \leq y \leq M(t)\}$ is continuous if m and M are continuous real-valued functions.

THEOREM 3.3.2. Under Assumptions A6, A7, A8, the mapping p_n described above satisfies Assumption A5.

Sketch of proof: Letting $(v_n, w_n) = p_n(y_n, u_n)$ as described above, it is easy to show that $|v_n(t_i) - y_{n,i}| = o(1)$ uniformly in i as $k \longrightarrow 0$ by examining the difference equation for $y_{n,i}$, the equation

$$v_n(t_{i+1}) = v_n(t_i) + \int_{t_i}^{t_{i+1}} s[t, v_n(t), u_{n,i}]\,dt$$

and using the continuity assumptions on $s(t, y, u)$. Since $d_n \geq d[y_{n,i}, Y(t_i)]$ tends to zero, we have $d[v_n(t_i), Y(t_i)] = o(1) + O(d_n)$ which, by the continuity of $Y(t)$, yields $d[v_n(t), Y(t)] = o(1)$. Similarly, we find $d[w_n(t), U(t)] = o(1)$. Writing

$$h(v_n, w_n) - h_n(y_n, u_n) = \sum \int_{t_i}^{t_{i+1}} \{c[t, v_n(t), u_{n,i}] - c[t_i, y_{n,i}, u_{n,i}]\} \, dt$$

and using the continuity property of $c(t, y, u)$, we find that

$$\lim_{n\to\infty} |h(v_n, w_n) - h_n(y_n, u_n)| = 0$$

and similarly for $g - g_n$. Q.E.D.

EXERCISE. Supply the details in the *Proof* of Theorem 3.3.2 above.

We remark that the estimates "$o(1)$" above are satisfactory for p_n, since we have no need for the actual bounds; dealing with r_n, however, we must have computable numbers d_n. Consider the definition now of an operator r_n, to be applied to $x^* = (y^*, u^*)$, the solution of the C-problem. Thus y^* satisfies

$$\dot{y}^*(t) = s[t, y^*(t), u^*(t)]$$

almost everywhere. Suppose for the moment we can define $u_n = r_n u^*$ via $u_{n,i} = u^*(t_i)$. Then $y_n = r_n y^*$ can be defined via

$$y_{n,i+1} = y_{n,i} + ks(t_i, y_{n,i}, u_{n,i}) \text{ for } 0 \le i \le n-1,\ y_{n,0} = y^*(0)$$

that is, so that y_n is a numerical solution of the differential equation for y^*; under suitable hypotheses we can then bound $y_{n,i} - y^*(t_i)$. If u^* is only measurable, we cannot estimate d_n but can only show that, for certain problems, there *exist* satisfactory d_n, using the techniques of Cullum (1969) as sketched in the first paragraph of this section. To derive computable d_n, we need more continuity assumptions on $u^*(t)$. Using these hypotheses we can bound $y^*_{n,i} - y^*(t_i)$ and hence bound d_n, while Assumption A8 is more than sufficient to guarantee $\lim_{u\to\infty} |h_n(x_n) - h(x^*)| = 0$.

ASSUMPTION A9. Assume that $s(t, y, u)$ is Lipschitz-continuous with respect to y uniformly in $(t, u) \in [0, 1] \times U^*$ and continuous in $(t, y, u) \in [0, 1] \times (-\infty, \infty) \times U^*$. Assume $u^*(t)$ is piecewise continuous, having only finitely many discontinuities, each of finite-jump type.

ASSUMPTION A10. Assume in addition to Assumption A9 that $s(t, y, u)$ is continuously differentiable with respect to t and u, and that $u^*(t)$ is piecewise continuously differentiable, both u^* and $\dot{u}^*$ having only finitely many discontinuities, each of finite-jump type.

THEOREM 3.3.3. Under Assumptions A8 and A9, r_n as described above satisfies Assumption A4, with $d_n = O[k + \omega(k)]$, where $\omega(k) \equiv \sup |s[t',$

$y, u^*(t')] - s[t'', y, u^*(t'')]$ with the supremum taken over all t', t'' with $0 \leq t' \leq t'' \leq 1, |t' - t''| \leq k$, t' and t'' in the same interval of continuity of u^*, and y in a certain bounded set R. If Assumption A10 holds, then $\omega(k) = O(k)$, and we may take the computable value $d_n = k^{1-\epsilon}$, $\epsilon > 0$.

Sketch of proof: The only real task is to bound $|y_{n,i} - y^*(t_i)|$. Were it not for the discontinuities in the equation, we could immediately write that $|y_{n,i} - y^*(t_i)| = O[k + \omega(k)]$ uniformly in i by the standard theory in Henrici (1962); it is trivial to generalize this to allow the discontinuities. Essentially the argument is as follows. Up to the first discontinuity τ_1 the $O[k + \omega(k)]$ result is valid. One can consider the calculation between τ_1 and the next discontinuity τ_2 as the solution of a new initial-value problem in which the initial data used in the numerical method—that is, $y^*(t_i)$ for the last $t_i \leq \tau_1$—are inaccurate of order $O[k + \omega(k)]$. Since the initial error propagates in a bounded fashion, the error on $[\tau_1, \tau_2]$ is also $O[k + \omega(k)]$. The argument proceeds in this manner throughout the finitely many discontinuities τ_j. Q.E.D.

EXERCISE. Supply the details for the *Proof* to Theorem 3.3.3, above.

The reader should note that we have only partly attained our goal of finding *computable* constants d_n. Our estimates—saying that we may take $d_n = O(k^{1-\epsilon})$, for example—only mean that the numerical method will thus work for sufficiently small k; we do not have a computable expression for d_n guaranteed to work for all k.

Although one would like to be able to prove convergence of the numerical solutions without the continuity requirements in Assumptions A9 and A10, this does not seem possible in general (for a special case, see Daniel [1970]); however, very broad classes of problems do have solutions satisfying A10, and one might even call this a typical situation. Thus the assumptions in A10 do not appear to be unreasonably strong. As a simple special case, the optimal-time problem for $\dot{y} = Ay + Bu$, with A and B constant, with $y(0) = y_0$ given, and with u restricted by $||u||_\infty \leq 1$, can be treated by making use of the classical theory of optimal-time processes; and it can be shown that, if a solution exists, it will be approximated by approximate solutions of the discretized problem with $k^{1-\epsilon}$ expanded constraint sets, extending slightly a result in Krasovskii (1957). More generally, under Assumptions A1–A10, we have proved that approximate solutions to a penalty-function form of the P-problem have weak limit points solving the C-problem.

Another approach for defining the mapping r_n without assuming the control $u^*(t)$ to be piecewise continuous is as follows (only the outline of the procedure is given). Suppose that, for each $\epsilon > 0$, $u^*(t)$ can be approximated by a continuous function $u_\epsilon(t)$ "nearly" satisfying the constraints—say,

$d[u_\epsilon(t), U(t)] \leq \delta_1(\epsilon)$ with $\delta_1(\epsilon) \to 0$ as $\epsilon \to 0$; and suppose that $y_\epsilon(t)$, defined as the solution to $\dot{y}_\epsilon = s(t, y_\epsilon, u_\epsilon)$, $y_\epsilon(0) = y^*(0)$, is also "near" the constraints—say, $d[y_\epsilon(t), Y(t)] \leq \delta_2(\epsilon)$, $d[y_\epsilon(1), Y_F] \leq \delta_2(\epsilon)$, with $\delta_2(\epsilon) \to 0$ as $\epsilon \to 0$—and "near" y^* so that

$$|h(y^*, u^*) - h(y_\epsilon, u_\epsilon)| \leq \delta_3(\epsilon) \qquad \text{with} \quad \delta_3(\epsilon) \to 0 \quad \text{as} \quad \epsilon \to 0$$

Pick n so large that the oscillation of u_ϵ over intervals of length k is less than ϵ, and define $u_{\epsilon,n}$ as the piecewise constant interpolant of u_ϵ at the points $0, k, 2k, \ldots$ and $y_{\epsilon,n}$ as the solution to $\dot{y}_{\epsilon,n} = s(t, y_{\epsilon,n}, u_{\epsilon,n})$, $y_{\epsilon,n}(0) = y^*(0)$. Again we can argue that $y_{\epsilon,n}$ and $u_{\epsilon,n}$ are "near" the constraint sets and $|h(y^*, u^*) - h(y_{\epsilon,n}, u_{\epsilon,n})| \leq \delta_4(\epsilon)$. For each n, let (z_n, w_n) be the solution (assuming it exists) of the original C-problem only with the control restricted to be constant on each interval $[t_i, t_{i+1})$, and define $(y_n, u_n) \equiv r_n(y^*, u^*)$ via $u_{n,i} = w_n(t_i)$, $y_{n,i+1} = y_{n,i} + ks[t_i, y_{n,i}, w_n(t_i)]$, $y_{n,0} = z_n(0)$. If, for example, $s(t, y, u)$ is (uniformly) Lipschitzian in y and t, then it is simple to see that $|y_{n,i} - z_n(t_i)| = O(k)$ uniformly in n and i.

EXERCISE. Prove that $|y_{n,i} - z_n(t_i)| = O(k)$ uniformly in n and i if $s(t, y, u)$ is uniformly Lipschitzian in y and t, as asserted in the preceding paragraph.

From this estimate for $y_n - z_n$ one can conclude that

$$|h(z_n, w_n) - h_n(y_n, u_n) \longrightarrow 0$$

Because of the minimal property of (z_n, w_n) and the fact that $(y_{\epsilon,n}, u_{\epsilon,n})$ is "near" the constraint set, one can conclude that

$$h(z_n, w_n) \leq h(y_{\epsilon,n}, u_{\epsilon,n}) + \delta_5(\epsilon)$$

Therefore, we can write $h(y^*, u^*) \leq h(z_n, w_n) \leq h(y_{\epsilon,n}, u_{\epsilon,n}) + \delta_5(\epsilon)$. Since

$$|h(y^*, u^*) - h(y_{\epsilon,n}, u_{\epsilon,n})| \leq \delta_4(\epsilon)$$

and

$$|h(z_n, w_n) - h_n[r_n(y^*, u^*)]| \longrightarrow 0$$

we conclude that $|h(y^*, u^*) - h_n[r_n(y^*, u^*)]| \to 0$. Thus Assumption A4 is satisfied for this r_n and d_n can be taken to be $k^{1-\epsilon}$ for any fixed $\epsilon > 0$.

EXERCISE. Consider the simpler C-problem in which $Y_I = \{y_0\}$, $Y_F = Y(t) = (-\infty, \infty)$, $u(t) = [-\alpha, \alpha]$ for some fixed α. Provide the detailed and precise hypotheses and arguments for the above construction of r_n. [For the solution of this problem, see Budak et al. (1968–69)].

3.4. CHEBYSHEV SOLUTION OF DIFFERENTIAL EQUATIONS

We wish to consider at this point a problem which can be examined best from the discretization viewpoint, although the theorems of Chapter 2 are not directly applicable. An attempt to apply the concepts of that chapter, however, will reveal the fundamental difficulties and research areas in the particular problem. This will show, as stated in Section 3.1, how the abstract discretization can be useful in guiding one's research.

Suppose one seeks to solve $Au = b$ where A is a uniformly elliptic linear (for simplicity here only) differential operator in two variables over a bounded domain D, under the condition $u = 0$ on Γ, the boundary of D, assumed to be sufficiently smooth; more general types of equations may also be treated by the method to be presented. A numerical method of recent popularity [Krabs (1963), Rosen (1968)], given a sequence of functions $\{\varphi_i\}$ satisfying the boundary data, consists in choosing numbers $a_{n,1}, \ldots, a_{n,n}$ to minimize

$$\max_{1 \leq j \leq M} \left| \left[A\left(\sum_{i=1}^{n} a_{n,i}\varphi_i \right) \right](x_j) - b(x_j) \right|$$

where the M points $\{x_j\}$ form a "grid" over D. Strictly for convenience we take $M = cn$ for fixed c (experience indicates that $c = 4$ is a good choice [Rosen (1968)]) and suppose that the grid is such that any point in D is at a distance of at most h_n from a grid point x_j. We wish to find conditions under which the miminizing point $u_n^* \equiv \sum_{i=1}^{n} a_{n,i}\varphi_i$ will converge, in some sense, to the solution u^* of our problem.

Since we seek to minimize a supremum norm, the norm must be defined; therefore let

$$E = \{u;\, u = 0 \text{ on } \Gamma, \text{ all partial derivatives of } u \text{ through second-order are continuous on } \bar{D} = D \cup \Gamma\}$$

For $u \in E$, let $\|u\| = \|u\|_\infty = \max_{x \in \bar{D}} |u(x)|$. Let

$$f(u) = \|Au - b\|_\infty$$

where we now need to assume that b is continuous and bounded on $\bar{D}$. Let E_n be that subset of E spanned by the functions $\varphi_1, \ldots, \varphi_n$, assumed to lie in E; let p_n be the identity mapping, and r_n be at the moment undefined. Define

$$f_n(u_n) = \|Au_n - b\|_{cn,\infty} = \max_{1 \leq i \leq cn} |[Au_n](x_i) - b(x_i)|$$

We now seek conditions for consistency. Consider condition 2 of Definition 2.2.2:

$$f(p_n u_n^*) - f_n(u_n^*) = \| Au_n^* - b \|_\infty - \| Au_n^* - b \|_{cn,\infty}$$

Since this quantity is always nonnegative, the requirement

$$\limsup_{n\to\infty} [f(p_n u_n^*) - f_n(u_n^*)] \leq 0$$

in fact demands convergence; in order to compare suprema over discrete and continuous sets, we need to know something about the growth of the functions $Au_n^* - b$ between grid points. Hence we now assume that $A\varphi_i$ satisfies a Lipschitz condition with Lipschitz constant λ_i (this restricts A somewhat also) and that b satisfies one with a constant λ_0. From this it follows that

$$|f(p_n u_n^*) - f_n(u_n^*)| \leq h_n \sum_{i=0}^{n} |a_{n,i}| |\lambda_i|$$

where $a_{n,0} = 1$. Thus we need next a growth condition on $\sum_{i=0}^{n} |a_{n,i}| |\lambda_i|$. For example, the following conditions would be sufficient:

1. that there exists a constant C such that $\sum_{i=0}^{n} |a_{n,i}| \leq C$ for all n; and
2. that $h_n \Lambda_n$ tends to zero, where $\Lambda_n = \max_{1 \leq i \leq n} \lambda_i$.

In practice, the Λ_n do in fact become large, while the restriction on the $a_{n,i}$ is easy to implement. In essence, the above restrictions are defining C_n—i.e.,

$$C_n = \left\{ u_n;\ \sum_{i=0}^{n} |a_{n,i}| \leq C \right\}$$

Next consider condition 1 of Definition 2.2.2, where r_n is to be defined. We require

$$\limsup_{n\to\infty} f_n(r_n u^*) \leq f(u^*)$$

Now

$$f_n(r_n u^*) \leq f(r_n u^*)$$

so we need only require that $\limsup_{n\to\infty} f(r_n u^*) \leq f(u^*)$; this is certainly true if $r_n u^*$ is an approximation method in which $Ar_n u^*$ converges uniformly to Au^*—if, for example, $r_n u^*$ and all its partial derivatives through second-order converge uniformly to those for u^*. Note that it is necessary to have $r_n u^*$ in C_n.

Under the above conditions, it follows in the same manner as Theorem 2.2.1 that

$$\lim_{n\to\infty} f_n(u_n^*) = \lim_{n\to\infty} f(p_n u_n^*) = f(u^*) = 0$$

where u^* solves $Au^* = b$ and lies in E; the conditions on weak sequential compactness and weak sequential lower semicontinuity are needed only to prove convergence for $p_n u_n^*$, a problem easily handled differently here. We know that

$$\| Au_n^* - b \|_\infty = f(p_n u_n^*)$$

converges to zero. By a simple use of the maximum principle [Protter–Weinberger (1967)], we deduce

$$\| u_n^* - u^* \|_\infty \leq \| Au_n^* - b \|_\infty \| w \|_\infty$$

where w solves $Aw = -1$ in D, $w = 0$ on Γ; therefore, u_n^* converges uniformly to the solution u^*.

EXERCISE. Provide the details for the above arguments showing that u_n^* converges uniformly to u^*.

The application of the theory in Chapter 2 to this problem indicates the type of approach necessary to prove convergence for this numerical method. We require: (1) smooth functions φ_i with Lipschitz constants λ_i for $A\varphi_i$ that do not grow too rapidly; (2) results from approximation theory that state that *if* one approximates functions b by combinations of functions $A\varphi_i$, the sums $\sum_{i=1}^{n} |a_{n,i}|$ remain bounded; and (3) results from approximation theory that state that functions b *can* be approximated by functions $A\varphi_i$. The requirements 1 and 3 here are probably less difficult; generalized Bessel-inequality results such as 2, however, are not known to this author for general cases. While numerical work with this method proceeds, theoretical results of the type suggested by Theorem 2.2.1 should and are being sought.

Using known general results from approximation theory [Rivlin–Cheney (1966)] comparing discrete and continuous approximations, we can avoid the questions of the growth of the $a_{n,i}$ and λ_i, although other problems arise. In particular, if b and the $\psi_i \equiv A\varphi_i$ are merely continuous, then there *exists* a sequence $\{h_n\}$ tending to zero so that, for the resulting discretization, $f(p_n u_n^*) - f_n(u_n^*)$ tends to zero, leading us to the uniform convergence of u_n^* to u^* as above. In general, however, we cannot give an explicit form for h_n; special results defining h_n can be given in one dimension in which the linear span of the ψ_i is the space of polynomials of a certain degree, but we are not aware of more widely applicable results in this direction.

EXAMPLE [Rosen (1968)]. The Chebyshev method described above can also be used on mildly nonlinear problems as well as linear ones, although the computation of the $a_{n,i}$ is then a *nonlinear* programming problem. We consider, for example, the approximate solution to

$$-\nabla^2 u + e^u = 50 \quad \text{in} \quad D$$
$$u = 0 \quad \text{on} \quad \partial D$$

where D is the unit square in two dimensions. Using 45 polynomials φ_i satisfying the boundary conditions exactly and using a grid of 225 points in the interior of D for computing the discrete maximum norm, a relative error *bound* of 0.0023 was computed, making use of a maximum principle for the bound. Using only 21 functions, the error bound increased to 0.021.

General references: Daniel (1968b), Rosen (1968).

3.5. CALCULUS OF VARIATIONS

We wish to consider now the standard problem in the calculus of variations for functions with given boundary values. For such problems over an arbitrary region in $\mathbb{R}^n$, it has been suggested [Greenspan (1967)] that a numerical solution be computed by minimizing a certain type of quadrature sum with derivatives in the integrand replaced by differences. The quadrature formula in two dimensions, for example, exactly integrates functions which are piecewise constant—in particular, constant over each component of a triangularization of the domain. In order to simplify the notation and eliminate some *minor* technical problems, we shall greatly specialize our analysis to the case of only one dimension. The techniques, assumptions, and results go over without essential change to rectangular domains in $\mathbb{R}^n$; we have not yet looked at the problem of arbitrary domains from the special viewpoint of discretizations. The space E, which we shall define, has of course some properties in $\mathbb{R}^n$ for $n > 1$ that are different from those for $n = 1$; in particular, weak convergence for $n > 1$ is rather "weaker." For a thorough analysis of the calculus of variations in $\mathbb{R}^n$, the reader is referred to Morrey (1966); relevant approximation concepts are in DiGuglielmo (1969).

Consider the problem of minimizing the functional

$$f(x) = \int_0^1 g(t, x, \dot{x})\, dt \qquad \text{subject to} \qquad x(0) = x(1) = 0$$

where $\dot{x} = dx/dt$. The following simple case of a general numerical method has been suggested [Greenspan (1967)]: minimize (or nearly minimize)

$$f_n(x_n) = \sum_{i=1}^{n} h_i g\left(t_{i-1}, x_{n,i-1}, \frac{x_{n,i} - x_{n,i-1}}{h_i}\right), \text{ subject to}$$

$$x_{n,0} = x_{n,n} = 0, \qquad h_i = t_i - t_{i-1}$$

where the minimization is over the set of values of $x_{n,1}, \ldots, x_{n,n-1}$; this method can be fitted neatly into the theory of Theorem 2.3.1. In Greenspan (1967), under the assumption that there exist unique minimizing points x^* for f (in $C^1[0, 1]$) and x_n^* for f_n satisfying the *spike condition*—

$$\left|\frac{x_{n,i}^* - x_{n,i-1}^*}{h_i}\right| \leq A$$

for some constant A independent of n—it was purportedly proved that $p_n x_n^*$, the piecewise linear interpolation to x_n^*, converges uniformly to x^*; because the author inadvertently left out an assumption guaranteeing a lower semi-continuity property for the functional f, the proof is in fact incorrect. However, as we shall show below by use of Theorem 2.3.1, the usual assumptions guaranteeing a unique minimizing point for f, in conjunction with an assumption guaranteeing the satisfaction of a type of spike condition, yield a convergence proof.

For convenience, let us take $h_i = h = 1/n$ for all i. For a fixed $p > 1$, let

$$E = \{x;\ x(0) = x(1) = 0,\ x \text{ is absolutely continuous on } [0, 1],\ \dot{x} \in L_p[0, 1]\}$$

For $x \in E$, let

$$\|x\| = \|\dot{x}\|_p = \left\{\int_0^1 |\dot{x}(t)|^p \, dt\right\}^{1/p}$$

For each n, let E_n be $(n - 1)$-dimensional Euclidean space where $x_n \in E_n$ has the norm

$$\|x_n\|_n = \left\{h \sum_{i=1}^{n} \left[\frac{|x_{n,i} - x_{n,i-1}|}{h}\right]^p\right\}^{1/p}$$

where $x_{n,0} = x_{n,n} = 0$ by definition. Let p_n be the mapping defined by piecewise linear joining of the values $x_{n,i}$ at $t_i = ih$, so $p_n x_n \in E$. Define the mapping r_n via $(r_n x)_i = x(t_i)$, $i = 1, \ldots, n - 1$.

We now make the standard type of assumption in the calculus of variations [Akhiezer (1962)] in order to guarantee the existence of a minimizing point for f. Note that E, as a closed linear subspace of $W_p^1(0, 1)$, is reflexive, and that weak convergence in E implies uniform convergence—that is, convergence in $C[0, 1]$—as noted in Section 1.3.

ASSUMPTIONS: A1. $g(t, x, w)$ is jointly continuous in its variables for $0 \leq t \leq 1$ and $-\infty \leq x, w \leq \infty$. A2. There exist constants a, b with $b > 0$ such that $g(t, x, w) \geq a + b|w|^p$ for all t in $[0, 1]$, x finite. A3. g is differentiably convex in w; i.e.,

$$g(t, x, w_1) - g(t, x, w_2) \geq (w_1 - w_2)g_w(t, x, w_2)$$

with g_w continuous in x, uniformly for (t, w) bounded.

PROPOSITION 3.5.1. The functional f is weakly sequentially lower semicontinuous on E, bounded below, and satisfies a T-condition.

Proof: For the last two assertions in this proposition, note that

$$f(x) = \int_0^1 g(t, x, \dot{x})\,dt \geq \int_0^1 [a + b|\dot{x}|^p]\,dt = a + b\|x\|^p$$

The proof of the weak sequential lower semicontinuity is straightforward using the convexity of g; details may be found in Akhiezer (1962), pp. 137–139. Q.E.D.

THEOREM 3.5.1. The discretization scheme defined above is stable and satisfies a uniform-growth condition.

Proof:

$$\begin{aligned}
\|p_n x_n\|^p &= \int_0^1 |(p_n x_n)^{\cdot}|^p\,dt \\
&= \sum_{i=1}^{n} \int_{t_{i-1}}^{t_i} |(p_n x_n)^{\cdot}|^p\,dt \\
&= h \sum_{i=1}^{n} \left[\frac{|x_{n,i} - x_{n,i-1}|}{h}\right]^p \\
&= \|x_n\|_n^p
\end{aligned}$$

proving stability. For the growth condition,

$$\begin{aligned}
f_n(x_n) &= h \sum_{i=1}^{n} g\left(t_{i-1}, x_{n,i-1}, \frac{x_{n,i} - x_{n,i-1}}{h}\right) \geq h \sum_{i=1}^{n} \left[a + b\left|\frac{x_{n,i} - x_{n,i-1}}{h}\right|^p\right] \\
&= a + b\|x_n\|_n^p
\end{aligned}$$

Q.E.D.

The only remaining ingredient for application of Theorem 2.3.1 is the consistency; in Greenspan (1967), the spike condition was needed for this. In our case, we must make the following assumptions.

ASSUMPTIONS: A4. Some solution x^* minimizing $f(x)$ lies in $C^1[0, 1]$; i.e., $\dot{x}^*$ is continuous. A5. There exist constants c and d and a continuous function $s(t, v)$ such that

$$|g(t_1, v_1, z) - g(t_2, v_2, z)| \leq (c + d|z|^p)|s(t_1, v_1) - s(t_2, v_2)|$$

where t_1, t_2 are aribtrary points in $[0, 1]$ and v_1, v_2, z are arbitrary real numbers.

Remarks. If

$$g(t, x, w) \equiv (w^2/2) + r(t, x)$$

then Assumption A5 is satisfied with $s \equiv r$. If

$$g(t, x, w) \equiv l(w)m(t, x) \quad \text{with} \quad |l(w)| \leq c + d|w|^p$$

then Assumption A5 is satisfied with $r \equiv m$; many actual problems are of the above types. Assumption A4 is probably superfluous in many cases.

THEOREM 3.5.2. The discretization described above is consistent.

Proof: For condition 1 of Definition 2.2.2 we prove

$$\lim_{n\to\infty} |f_n(r_n x^*) - f(x^*)| = 0$$

Since, by assumption, x^* is in $C^1[0, 1]$, given ϵ, for sufficiently large n,

$$|x^*(t_{i-1}) - x^*(t)| < \epsilon \quad \text{and} \quad \left|\dot{x}^*(t) - \frac{x_i^* - x_{i-1}^*}{h}\right| < \epsilon \quad \text{for} \quad t_{i-1} \leq t \leq t_i$$

Thus,

$$|f(x^*) - f_n(r_n x^*)| \leq \sum_{i=1}^{n} \int_{t_{i-1}}^{t_i} \left| g(t, x^*, \dot{x}^*) - g\left(t_{i-1}, x_{i-1}^*, \frac{x_i^* - x_{i-1}^*}{h}\right)\right| dt$$

But, by uniform continuity of g, given $\delta > 0$ there exists $\epsilon > 0$ and then N such that $n > N$ implies

$$|f(x^*) - f_n(r_n x^*)| \leq \sum_{i=1}^{n} \int_{t_{i-1}}^{t_i} \delta \, dt = \delta$$

Since $\delta > 0$ is arbitrary, condition 1 is proved. For condition 2 of Definition 2.2.2, we show that $\lim_{n\to\infty} |f(p_n x_n) - f_n(x_n)| = 0$ if $\| p_n x_n \|$ is bounded:

$$|f(p_n x_n) - f_n(x_n)| \leq h \sum_{i=1}^{n} \int_0^1 \left| g\left[t_{i-1} + \alpha h, (1-\alpha)x_{n,i-1} + \alpha x_{n,i}, \frac{x_{n,i} - x_{n,i-1}}{h}\right] \right.$$

$$\left. - g\left[t_{i-1}, x_{n,i-1}, \frac{x_{n,i} - x_{n,i-1}}{h}\right] \right| d\alpha$$

$$\leq h \sum_{i=1}^{n} \int_0^1 \left(c + d\left|\frac{x_{n,i} - x_{n,i-1}}{h}\right|^p\right)$$

$$\times |s(t_{i-1} + \alpha h, (1-\alpha)x_{n,i-1} + \alpha x_{n,i}) - s(t_{i-1}, x_{n,i-1})| \, d\alpha$$

Now, $||x_n||_n = ||p_n x_n||$ is bounded, $|x_{n,i}|$ is bounded, and

$$|x_{n,i} - x_{n,i-1}| \leq h^{1-1/p} ||x_n||_n$$

Thus, using the uniform continuity of $s(t, x)$, given $\epsilon > 0$, there exists N such that $n > N$ implies

$$|f(p_n x_n) - f_n(x_n)| \leq h \sum_{i=1}^{n} \int_0^1 \left(c + d\left|\frac{x_{n,i} - x_{n,i-1}}{h}\right|^p\right) \epsilon \, d\alpha = \epsilon[c + d||x_n||_n^p]$$

Since $\epsilon > 0$ is arbitrary, condition 2 follows. Q.E.D.

EXERCISE. Show that $|x_{n,i}|$ is bounded independently of n, i as asserted the *Proof* of Theorem 3.5.2.

We now can state the following theorem which follows immediately from Theorem 2.3.1 and the above theorems.

THEOREM 3.5.3. Let Assumptions A1 — A5 be valid and let the discretization method described above be used. Then all weak limit points of $p_n x_n^*$, at least one of which exists, minimize f. If the solution x^* is unique, then, in particular, $p_n x_n^*$ converges uniformly to x^* and the derivatives converge L_p-weakly.

EXAMPLE [Greenspan (1965)]. Consider minimizing

$$\int_0^1 |x|(1 + \dot{x}^2)^{1/2} \, dt, \quad \text{subject to} \quad x(0) = 1, \quad x(1) = \cosh 1$$

having solution $x(t) = \cosh t$. Using $h = 0.2$, a maximum error of 0.046 is found, while for $h = 0.01$ the error is 0.0015.

For similar results, see Simpson (1968, 1969).

3.6. TWO-POINT BOUNDARY-VALUE PROBLEMS

The problem discussed in the previous section is of course essentially a two-point boundary-value problem for a second-order ordinary differential equation; the method described is only one of many possible for use on this problem. Another recent method of great interest is the application of the Ritz method to this problem, using certain special classes of functions as basis functions. In Section 3.7 we shall examine the general Ritz method, but in this section we wish to look at the more special problem indicated above. For clarity we shall consider only simple boundary conditions, although more complex ones can be treated [Ciarlet et al. (1968a, b)]. The method has been thoroughly analyzed [Ciarlet (1966), Ciarlet et al. (1967)] for solving

$$\sum_{j=0}^{n} (-1)^{j+1} D^j[q_j(t) D^j x(t)] = g[t, x(t)], \quad t \in (0, 1)$$
$$D^k x(0) = D^k x(1) = 0, \quad k = 0, 1, \ldots, n-1$$

where $Dy \equiv dy/dt$. The results in this general case, if $q_n(t) \geq \epsilon > 0$, are more complicated to state, but just the same as those for the equation

$$D^2 x(t) = g[t, x(t)], \quad t \in (0, 1) \qquad x(0) = x(1) = 0 \tag{3.6.1}$$

that is, for $n = 1$, $q_0(t) \equiv 0$, $q_1(t) \equiv 1$; therefore, we shall present only this simpler but sufficiently representative problem.

Let the Hilbert space $E \equiv {}_0W_2^1 \equiv \{x; x$ is absolutely continuous, $\dot{x} \in L_2(0, 1)$, $x(0) = x(1) = 0\}$, and, for $x, y \in E$, define

$$\langle x, y \rangle = \int_0^1 Dx(t)\, Dy(t)\, dt$$

Assume that $g(t, x)$ is continuous in (t, x) in $[0, 1) \times (-\infty, \infty)$, and satisfies

(1) $$\frac{g(t, x) - g(t, y)}{x - y} \geq \gamma > -\pi^2 \qquad \text{if} \quad x \neq y$$

(2) $$\frac{g(t, x) - g(t, y)}{x - y} \leq M(c) < \infty \qquad \text{if} \quad |x| \leq c, \quad |y| \leq c$$

Define the functional

$$f(x) \equiv \int_0^1 \left\{ \tfrac{1}{2}[Dx(t)]^2 + \int_0^{x(t)} g(t, z)\, dz \right\} dt$$

It is easy to deduce that, if $x^*(t)$ is a classical solution of Equation 3.6.1, then x^* minimizes f over E. Clearly also x^* is the unique minimizing point, since f is convex and, in particular,

$$f(x + y) \geq f(x) + \int_0^1 \{[Dx(t)][Dy(t)] + y(t) g[t, x(t)]\}\, dt + (\gamma + \pi^2) \int_0^1 y^2(t)\, dt$$

which implies

$$f(x^* + y) \geq f(x^*) + (\gamma + \pi^2) \int_0^1 y^2(t)\, dt$$

as we have seen before. Moreover, if S_M is a subspace of dimension M spanned by the functions $\varphi_1, \ldots, \varphi_M$, then there exists a unique element ϕ_M in S_M minimizing f over S_M,

$$\phi_M = \sum_{i=1}^{M} a_i \varphi_i$$

which is also the unique solution of

$$\frac{\partial f\left(\sum_{i=1}^{M} a_i \varphi_i\right)}{\partial a_i} = 0, \qquad i = 1, \ldots, M$$

that is,

$$Ba + G(a) = 0 \tag{3.6.2}$$

where $a = (a_1, \ldots, a_M)^T$, B is the matrix $B = ((B_{ij}))$

$$B_{ij} \equiv \langle \varphi_i, \varphi_j \rangle = \int_0^1 [D\varphi_i(t)][D\varphi_j(t)]\, dt$$

$$G(a) = [G_1(a), \ldots, G_M(a)]^T$$

$$G_i(a) = \int_0^1 g\left[t, \sum_{j=1}^{M} a_j \varphi_j(t)\right] \varphi_i(t)\, dt$$

For various kinds of subspaces S_M, bounds on the error between ϕ_M and x^* have been computed; the basic argument for obtaining the bound is simple. If we write $\nabla f(x) = J(x)$, then since x^* minimizes f on E we have $J(x^*) = 0$, while $\langle J(\phi_M), \varphi_i \rangle = 0$ since ϕ_M minimizes f over S_M. Thus

$$0 = \langle J(x^*) - J(\phi_M), \varphi_i \rangle = \langle J'_{x_0}(x^* - \phi_M), \varphi_i \rangle$$

for some fixed x_0. Defining $[x, y] \equiv \langle J'_{x_0}(x), y \rangle$ as a new inner product, we see that ϕ_M is the closest point to x^* in S_M in the sense of this inner product. Thus any theorems about how well x^* can be approximated by elements of S_M can be used to lead to statements about the error $x^* - \phi_M$ in various norms.

Much of the theory has been developed for the case of S_M being various "piecewise polynomial" subspaces, making use of the well-developed theory of *spline* and polynomial approximation. For example, let P denote the partition $0 = t_0 < t_1 < \cdots < t_{N+1} = 1$ of $[0, 1]$. For $m \geq 1$, we define a class of splines $H_0^m(P) = \{\varphi(t);\ \varphi$ is in $C^{m-1}[0, 1]$, $\varphi(0) = \varphi(1) = 0$ and φ is a polynomial of degree at most $2m - 1$ on $[t_i, t_{i+1}]$ for $0 \leq i \leq N\}$. This space $H_0^m(P)$ is spanned by the $m(N + 2) - 2$ functions $S_{i,k}(t)$ for $1 \leq i \leq N$, $0 \leq k \leq m - 1$, and for $i = 0, N + 1$, $1 \leq k \leq m - 1$, where

$$D^l S_{i,k}(t_j) = \delta_{i,j}\delta_{k,l} \quad \text{for} \quad 0 \leq l \leq m - 1$$

The functions $S_{i,k}(t)$ are zero except in $[t_{i-1}, t_{i+1}]$. For example, with $m = 1$, $H_0^1(P)$ is spanned by the N functions $S_{i,0}$, $1 \leq i \leq N$, where $S_{i,0}(t)$ is given by the roof function (Figure 3.1).

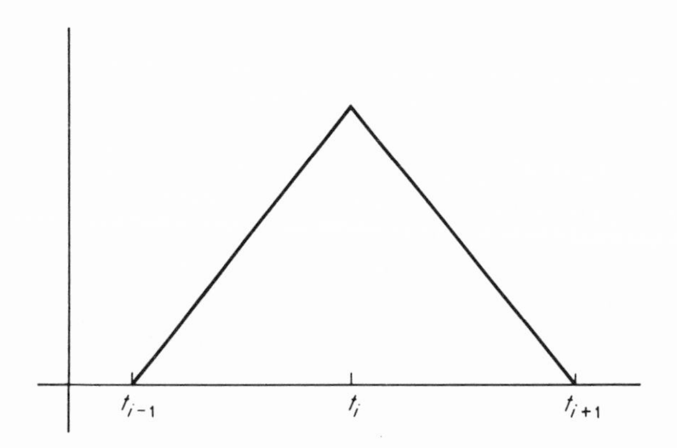

Figure 3.1

EXERCISE. Find the basis functions for $H_0^2(P)$.

By using known results about approximation (in fact, interpolation) by elements of $H_0^m(P)$, we can give bounds for $\phi_M - x^*$, as described above, where $M = m(N + 2) - 2$. For example [Ciarlet et al. (1967)], if

$$|P| \equiv \max_{0 \leq i \leq N} |t_{i+1} - t_i|$$

then if $x^* \in C^q[0, 1]$ with $q \geq 2m$, then there exists a constant K such that

$$\|D^k(\phi_M - x^*)\|_\infty \leq K\|D^{2m}x^*\|_\infty |P|^{2m-1}, \qquad k = 0, 1$$

where $\|u\|_\infty = \operatorname*{ess\,sup}_{0 \leq t \leq 1} |u(t)|$ for $u \in L_\infty(0, 1)$. Thus, if $x^* \in C^2[0, 1]$, we can use H_0^1 as our subspace and find error bounds of order $|P|$. If $x^* \in C^4[0, 1]$ we can use H_0^2 and find bounds of order $|P|^3$. In fact, by more subtle arguments [Perrin–Price–Varga (1969)], one can show that the order of convergence for H_0^1 approximation is actually $|P|^2$.

In a practical sense, however, the above results are not meaningful unless one can compute ϕ_M; that is, solve

$$Ba + G(a) = 0$$

For the type of subspaces we are considering, the matrix B can be assumed to be known exactly, since it is computed by integration using polynomials; using H_0^m spaces, the matrix in fact is a band matrix. The operator $G(a)$, however, involves integration of $g\left[t, \sum_{i=1}^{M} a_i\varphi_i(t)\right]$, which we cannot perform exactly; the use of a quadrature formula gives us a computable method [Herbold (1968), Herbold et al. (1969)]. Suppose we use a quadrature formula

$$\int_{y_0}^{y_k} s(y)\,dy \doteq \sum_{i=0}^{k} \alpha_i s(y_i) \tag{3.6.3}$$

with error given by

$$K_1\left(\frac{y_k - y_0}{k}\right)^{k_0+1} s^{(k_0)}(\xi)$$

as usual. Given the partition P: $0 = t_0 < t_1 < \cdots < t_{N+1} = 1$, if we write

$$\int_0^1 s(t)\varphi_i(t)\,dt = \sum_{i=0}^{N} \int_{t_i}^{t_{i+1}} s(t)\varphi_i(t)\,dt$$

and apply the quadrature formula to each subinterval, we obtain a composite quadrature formula

$$\int_0^1 s(t)\varphi_i(t)\,dt \simeq \sum_{j=0}^{M_0} \beta_j s(t'_j)\varphi_i(t'_j)$$

where $M_0 = k(N + 1)$ and the β_j and t'_j are obtainable from the α_j and t_j.

EXERCISE. Find explicit expressions for the β_j and t'_j in the composite quadrature formula above.

If we use the above sum to approximate the operator $G(a)$, we get

$$\bar{G}_k(a) = [\bar{G}_{k,1}(a), \ldots, \bar{G}_{k,M}(a)]^T$$

$$\bar{G}_{k,i}(a) = \sum_{j=0}^{k(N+1)} \beta_j g[t'_j, \sum_{l=1}^{M} a_l\varphi_l(t'_j)]\varphi_i(t'_j)$$

and we now can solve

$$Ba + \bar{G}_k(a) = 0 \tag{3.6.4}$$

numerically. This, however, is the gradient equation for the functional

$$f_k(a) = \int_0^1 \tfrac{1}{2}\left\{D\left[\sum_{j=1}^{M} a_j\varphi_j(t)\right]\right\}^2 dt + \sum_{j=0}^{k(N+1)} \beta_j \int^{\sum_{i=1}^{M} a_i\varphi_i(t_j')} g(t_j', \eta)\, d\eta$$

EXERCISE. Prove that Equation 3.6.4 above is equivalent to $\nabla f_k(a) = 0$, where f_k is as defined above.

If one assumes that the quadrature scheme of Equation 3.6.1 has $\alpha_i \geq 0$, $\sum_{i=0}^{k} \alpha_i = y_k - y_0$ (which implies $\beta_j \geq 0$ and $\sum_{j=0}^{k(N+1)} \beta_j = 1$), and that

$$\sum_{j=0}^{k(N+1)} \beta_j \varphi_i(t_j')\varphi_l(t_j') = \int_0^1 \varphi_i(t)\varphi_l(t)\, dt \qquad \text{for} \quad 1 \leq i \leq l \leq M$$

—that is, our formula computes $\int_0^1 \varphi_i(t)\varphi_l(t)\, dt$ exactly—then one can show that there is a unique $a^* \equiv a_k^* = (a_{k,1}^*, \ldots, a_{k,M}^*)^T$ minimizing $f_k(a)$ for each k and solving Equation 3.6.4. The condition on the positivity of the weights is valid, of course, for all Gauss formulas and for L-point Newton–Cotes formulas with $2 \leq L \leq 8$, while the condition on integration of the *polynomial* $\varphi_i\varphi_l$ requires that the quadrature formula be of a certain degree.

Thus what we now have is a discretization scheme. Rather than minimize $f(x)$ over E, we now minimize $f_k(a)$ for $a \in \mathbb{R}^M$; the various parameters M, N, and k must of course be related in some way. We want the vector $\phi_{M,k}^* = \sum_{i=1}^{M} a_{k,i}^*\varphi_i$ to be near x^* and infact to be as accurate as ϕ_M itself. If for some set of spaces S_M we have the error bound for $x^* - \phi_M$ as a function $e_M(x^*)$, we shall say that the quadrature scheme is *compatible* if the error $\phi_{M,k}^* - x^*$ is also bounded by const$\cdot e_M(x^*)$. From the discretization viewpoint, we can prove the existence of the mapping r_n by using interpolation of x^* in S_M, and we can take p_n to be the obvious embedding operator. We now state what is the final result for use of the space $H_0^m(P)$.

PROPOSITION 3.6.1 [Herbold (1968)]. Suppose we use the subspaces $H_0^m(P_N)$ for a partition P_N: $0 = t_0 < t_1 < \cdots < t_{N+1} = 1$, a subspace of dimension $m(N+2) - 2$, where the partition P_N saitsfies $|P_N| \leq L \min_{0 \leq i \leq N} |t_{i+1} - t_i|$ for all N. Suppose $g(t, x)$ is so smooth that for any $\varphi \in H_0^m(P_N)$, $D^s g[t, \varphi(t)]$ is continuous on each interval of the partition for $0 \leq s \leq k_0$, where k_0 describes the accuracy of the quadrature formula of Equation 3.6.3 and satisfies $k_0 \geq 4m - 1$. Suppose that the weights α_i in the quadrature formula satisfy $\alpha_i \geq 0, \sum_{i=0}^{k} \alpha_i = y_k - y_0$. Suppose x^* solving the original problem is in $C^{2m}[0, 1]$. Then $\phi_M = \phi_{m(N+2)-2}$ minimizing f over H_0^m and $\phi_{M,k}^*$ minimizing f_k are related by

$$\|D^l(\phi_M - \phi^*_{M,k})\|_\infty = O(|P_N|^{2m}), \qquad l = 0, 1$$

and thus the quadrature scheme is compatible and

$$\|D^l(x^* - \phi^*_{M,k})\|_\infty = O(|P_N|^{2m-1}), \qquad l = 0, 1$$

To be more specific, consider use of $H^1_0(P_N)$; for a quadrature scheme one could use a two-point Gaussian scheme, but we consider

$$\int_{y_0}^{y_2} s(t)\,dt \sim \frac{y_2 - y_0}{6}[s(y_0) + 4s(y_1) + s(y_2)], \qquad k_0 = 4$$

Since $k_0 \geq m - 1$, we deduce

$$\|D^l(\phi_M - \phi^*_{M,2})\|_\infty = O(|P_N|^2), \qquad l = 0, 1$$

For the special case of H^1_0 we know that the error in $\phi_M - x^*$ is also $O(|P_N|^2)$, so we conclude that, if $x^* \in C^2[0, 1]$,

$$\|D^l(x^* - \phi^*_{M,2})\|_\infty = O(|P_N|^2), \qquad l = 0, 1 \quad N \longrightarrow \infty$$

As a further example, the four-point Gaussian scheme, with $k_0 = 8$, is compatible with $H^2_0(P_N)$, so that the error using this subspace and quadrature formula, if $x^* \in C^4[0, 1]$, is of order $|P_N|^3$.

As a concrete example, consider

$$D^2x(t) = e^{x(t)}, \qquad x(0) = x(1) = 0$$

to be solved using $H^2_0(P_N)$, P_N: $t_i = ih$, $h = 1/(N + 1)$, and the compatible four-point Gaussian integration scheme. The errors $\|x^* - \phi^*_{M,2}\|_\infty$ are: 3.13×10^{-5} at $h = \frac{1}{3}$, 4.40×10^{-6} at $h = \frac{1}{5}$, and 7.15×10^{-7} at $h = \frac{1}{8}$ [Ciarlet et al. (1967), Herbold (1968)]. In this special case, one can show that $\|x^* - \phi^*_{M,2}\|_\infty \leq Kh^{7/2}$ [Perrin–Price–Varga (1969)].

General references: Herbold (1968), Simpson (1968).

3.7. THE RITZ METHOD

The method discussed in the previous section was simply a special case of the general *Ritz method.* Suppose we wish to minimize a functional $f(x)$ over a Hilbert space E. Suppose that for each n we have a finite dimensional subspace E_n of E with the property that for each x in E (actually, we need this only for the point x^* minimizing f),

$$\lim_{n\to\infty} d(x, E_n) = 0$$

where $d(x, E_n) = \min_{x_n \in E_n} \|x - x_n\|$. We can write "min" rather than "inf" above since in a Hilbert space the distance to a closed linear subspace is always attained, and uniquely so. For each n, we then find x_n^* minimizing f over E_n and hope that x_n^* converges to x^* in some sense.

We can describe this easily as a discretization method. For each n let $f_n \equiv f$ and let p_n be the identity mapping; let r_n be the best approximation mapping—that is, $\|x - r_n x\| = \min_{x_n \in E_n} \|x - x_n\|$. If f is norm-upper semi-continuous, then this discretization is consistent. Employing the conditions of Definition 2.2.2 to check this, for condition 1 we need $\limsup_{n\to\infty} f(r_n x^*) \geq f(x^*)$, which follows directly from the norm-upper semicontinuity, the definition of r_n, and the assumption that $d(x^*, E_n)$ approaches zero. For condition 2, we have $f_n(x_n^*) - f(p_n x_n^*) = f(x_n^*) - f(x_n^*) = 0$; the other conditions are irrelevant here. Thus by a simple modification in Theorem 2.3.1, we have the following; essentially the same theorem is valid for minimization over a set C, when the Ritz problem is then solved over $C \cap E_n$.

PROPOSITION 3.7.1. Let f be a weakly sequentially lower semicontinuous, norm-upper semicontinuous functional on the Hilbert space E and let f satisfy a T-property. Let $\{E_n\}$ be a sequence of closed linear subspaces such that $\lim_{n\to\infty} d(x^*, E_n) = 0$, and for each n let x_n^* satisfy

$$f(x_n^*) \leq f(x_n) + \epsilon_n \quad \text{for all} \quad x_n \in E_n$$

with $\lim_{n\to\infty} \epsilon_n = 0$. Then all weak limit points of x_n^*, at least one of which exists, minimize f over E, and $\lim_{n\to\infty} f(x_n^*) = f(x^*)$.

EXERCISE. Give a rigorous *Proof* for Proposition 3.7.1.

As mentioned before and shown in a previous section, if f satisfies some type of a uniform-convexity assumption, then x^* is unique and $x_n^* \to x^*$. As also shown in a previous section, in practice one cannot compute f precisely but must use some approximation to it; for the particular example of the preceding section—namely, two-point boundary-value problems—we saw that this still could lead to satisfactory results under suitable hypotheses. Clearly we could consider this problem in greater generality via the discretization viewpoint; we prefer to leave this as an exercise and look briefly instead at some known results [Mikhlin–Smolitskiy (1967)] in this direction for the special case in which f is a quadratic functional.

Suppose we seek to solve the equation

$$Ax = k$$

where A is a bounded, positive-definite, self-adjoint linear operator in a

Hilbert space E; this equation has a unique solution, x^*, which clearly must also be the unique point minimizing the functional

$$f(x) = \tfrac{1}{2}\langle Ax, x\rangle - \langle x, k\rangle$$

over E. Since $f''_x = A$ for all x, f is convex and in fact

$$\begin{aligned} f(x+h) &= f(x) + \langle \nabla f(x), h\rangle + \tfrac{1}{2}\langle Ah, h\rangle \\ &\geq f(x) + \langle \nabla f(x), h\rangle + \tfrac{1}{2}m\|h\|^2 \end{aligned}$$

where $0 < mI \leq A$ and $\nabla f(x) = Ax - k$. Thus if $\lim_{n\to\infty} f(x_n) = f(x^*)$, we have

$$f(x_n) - f(x^*) \geq \langle Ax^* - k, x_n - x^*\rangle + \frac{m}{2}\|x_n - x^*\|^2$$

which implies $x_n \longrightarrow x^*$. Thus if we use the Ritz method on this f, we find that $x_n^* \longrightarrow x^*$.

We suppose that, for the Ritz method, E_n is, for each n, the linear subspace spanned by $\varphi_1, \ldots, \varphi_n$ where $\{\varphi_1, \ldots\}$ is a complete basis for E—that is, a set of linearly independent elements with $d(x, E_n) \longrightarrow 0$ for each $x \in E$. Then minimizing f over E_n is precisely equivalent to solving

$$A_n x_n^* = k_n$$

where $k_n \in \mathbb{R}^n$, $k_n = (\langle\varphi_1, k\rangle, \langle\varphi_2, k\rangle, \ldots, \langle\varphi_n, k\rangle)^T$ and A_n is an $n \times n$ matrix, $A_n = ((A_{n,i,j}))$,

$$A_{n,i,j} = \langle A\varphi_i, \varphi_j\rangle$$

EXERCISE. Prove that minimizing f over E_n is equivalent to solving $A_n x_n^* = k_n$ for A_n and k_n as described above.

Since the $\{\varphi_i\}$ are linearly independent, there is a unique solution $x_n^* = \sum_{i=1}^{n} a_{n,i}\varphi_i$ for each n. We recognize however that the vector k_n and matrix A_n will not be computed exactly but will involve some errors. Thus we shall compute $A'_n = A_n + \Gamma_n$, $k'_n = k_n + \delta_n$ where $|\Gamma_n|_n$ and $|\delta_n|_n$ are assumed to be small, and $|\cdot|_n$ denotes a standard norm on $\mathbb{R}^n$ (and its induced matrix norm). We wish to see how this affects the computed x_n^*.

DEFINITION 3.7.1. Let $\{\psi_1, \ldots\}$ be a finite or infinite sequence in a Hilbert space E. The sequence is called *minimal* if the removal of each single element ψ_i restricts the subspace spanned by the sequence.

EXERCISE. Show that a finite set of linearly independent elements is minimal and that an infinite set of orthonormal elements is minimal.

DEFINITION 3.7.2. An infinite sequence $\{\psi_1, \ldots\}$ in a Hilbert space E with inner product $(\cdot, \cdot)$ is called *strongly minimal* if the smallest eigenvalue of the matrix

$$M_n \equiv (((\psi_i, \psi_j)))_{i,j=1,\ldots,n}$$

is bounded below by a positive number independent of n.

EXERCISE. Show that any orthonormal sequence is strongly minimal and that a strongly minimal sequence is a minimal sequence.

The following result is known [Mikhlin-Smolitskiy (1967)].

PROPOSITION 3.7.2. If the coordinate system $\{\varphi_1, \ldots\}$ is strongly minimal in E, then the solution by the Ritz method is stable under small variations in the matrix and right-hand side; that is, if $A'_n x'_n = k'_n$, $A'_n = A_n + \Gamma_n$, and $k'_n = k_n + \delta_n$, then for $|\Gamma_n|_n$ and $|\delta_n|_n$ sufficiently small there exist constants c_1 and c_2 independent of n such that

$$\|x'_n - x^*_n\| \leq c_1 |\Gamma_n|_n + c_2 |\delta_n|_n$$

We still are not being realistic computationally, however, since we are only considering x'_n, the *exact* solution of the perturbed equations. In attempting to solve the linear system $A'_n x'_n = k'_n$ we shall of course make further errors. The size of these errors, roughly speaking, varies directly with the *condition number* of the matrix A'_n—say, the ratio of its largest to smallest eigenvalues. Even for a general strongly minimal system, it is possible for this ratio to grow without bound as n increases. One can say the following, however, for strongly minimal sets of a special nature [Mikhlin–Smolitskiy (1967)]:

PROPOSITION 3.7.3. Let B be a positive-definite self-adjoint bounded linear operator on E which satisfies

$$m_1\langle Ax, x\rangle \leq \langle Bx, x\rangle \leq M_1\langle Ax, x\rangle \quad \text{for all} \quad x \in E$$

with $m_1 > 0$, $M_1 < \infty$. Suppose the complete basis $\{\varphi_1, \ldots\}$ satisfies

$$\langle B\varphi_i, \varphi_j\rangle = \begin{cases} 1 & \text{if} \quad i = j \\ 0 & \text{if} \quad i \neq j, \quad i, j \geq 1 \end{cases}$$

Then the basis is strongly minimal and moreover the condition number of A_n is bounded by M_1/m_1 and thus that of A'_n is uniformly bounded for $|\Gamma_n|_n$ and $|\delta_n|_n$ sufficiently small. In particular, these statements hold if $\{\varphi_1, \ldots\}$ is a complete orthonormal basis in E.

The Ritz method has been very popular for use in solving certain differential equations of physical interest, and much theory has been developed in this area. For more details and excellent examples the reader is referred to Mikhlin–Smolitskiy (1967).

A special feature of the finite dimensional subspaces E_n, namely that E_n is the span of $\varphi_1, \ldots, \varphi_n$, allowed us to derive the special results above; often, however, one does not have such expanding subspaces. For example, in $\mathbb{R}^1$, if E_n is the set of piecewise linear functions on [0, 1] with nodes at i/n, $i = 1, 2, \ldots, n - 1$, we have no such expanding basis; another feature in this case makes analysis easy as we shall now see. More generally, in $\mathbb{R}^l$, let $\varphi(x)$ be a function with compact support and define

$$\varphi_j^h(x) \equiv h^{-l/2}\varphi\left(\frac{x}{h} - j\right)$$

for x in $\mathbb{R}^l$ and l-integers j; if we let E_n (for $h = 1/n$) be the space of linear combinations of these functions, we have the *finite-element method.* Some steps have been taken to analyze this very general method [Fix–Strang (1969)]. For example, on the sample problem

$$-u_{x_1x_1} - u_{x_2x_2} + u = f(x_1, x_2)$$

in $\mathbb{R}^2$, the relevant square matrices important in the Ritz method (finite-element method) have uniformly bounded inverses (in the l_2-norm) if and only if there is no θ_0 in $\mathbb{R}^2$ such that $\hat{\varphi}(2\pi j + \theta_0) = 0$ for all 2-integers j, where $\hat{\varphi}(\xi)$ is the *Fourier transform*

$$\hat{\varphi}(\xi) \equiv \int_{\mathbb{R}^2} \varphi(x) e^{-i\xi^T x}\, dx$$

Moreover, the resulting numerical method is convergent if and only if for some integer $p \geq 1$, $\hat{\varphi}(0) \neq 0$ but $\hat{\varphi}$ has zeros of order at least $p + 1$ at $\xi = 2\pi j$ for all other 2-integers j. More widely applicable theory is under development.

EXERCISE. In $\mathbb{R}^1$ rather than $\mathbb{R}^2$, find some functions φ having Fourier transforms satisfying the above necessary-and-sufficient condition for convergence.

4 GENERAL BANACH-SPACE METHODS OF GRADIENT TYPE

4.1. INTRODUCTION

We now wish to consider iterative methods for minimizing a functional f in some real Banach space E; primarily, we shall be concerned with the unconstrained problem—that is, minimization over all of E—but we shall also briefly consider methods for the constrained problem when they are natural extensions of earlier methods. If f is differentiable, then from the formula

$$\frac{d}{dt} f(x + tp)\bigg|_{t=0} = \langle p, \nabla f(x) \rangle$$

we see that f is instantaneously decreasing most rapidly in the *direction* p (that is, with $\|p\| = 1$) if

$$\langle p, \nabla f(x) \rangle = -\|\nabla f(x)\|$$

In a Hilbert space this gives

$$p = \frac{-\nabla f(x)}{\|\nabla f(x)\|}$$

the direction of *steepest descent* [Cauchy (1847), Curry (1944)]. More generally [Altman (1966a, b)], we consider a steepest-descent direction to be any direction $p \in E$, $\|p\| = 1$, such that $\langle p, \nabla f(x) \rangle = -\|\nabla f(x)\|$. If the unit sphere in E is strictly convex—that is, if $\|x\| = \|y\| = 1$ and $0 < \lambda < 1$ imply $\|\lambda x + (1 - \lambda)y\| < 1$—then such a direction p is unique.

The function f of course instantaneously decreases in any direction p satisfying $\langle p, \nabla f(x) \rangle \leq 0$. We shall consider iterative methods which, at each point x_n in the iterative sequence, provide such a direction $p_n(x_n)$

along which we move to the next point $x_{n+1} = x_n + t_n p_n$ where the distance t_n of movement must be expeditiously chosen. We must be sure that the directions do not become nearly orthogonal to $\nabla f(x)$ too rapidly.

DEFINITION 4.1.1. A sequence of vectors $p_n(x_n)$ will be called *admissible* if and only if $\langle p_n(x_n), \nabla f(x_n)\rangle \leq 0$, and $\| \nabla f(x_n) \| \longrightarrow 0$ whenever

$$\left\langle \frac{p_n(x_n)}{\| p_n(x_n) \|}, \nabla f(x_n) \right\rangle \longrightarrow 0$$

For example a sequence of steepest-descent directions is admissible, as is a sequence generated by $p_n(x_n) = -B\nabla f(x_n)$ where $B: E^* \longrightarrow E$ satisfies $\langle Bc, c\rangle \geq m \| c \|^2, m > 0$, for all $c \in E^*$. We shall be able to analyze the iteration $x_{n+1} = x_n + t_n p_n(x_n)$ for admissible direction sequences $\{p_n(x_n)\}$ and various methods of choosing $t_n \geq 0$.

4.2. CRITICIZING SEQUENCES AND CONVERGENCE IN GENERAL

Throughout this chapter we shall denote by $W(x_0)$ the following set:

$W(x_0) \equiv$ the intersection of all norm-closed convex sets containing $L(x_0) \equiv \{x; f(x) \leq f(x_0)\}$ as a subset

Thus $W(x_0)$ is the *closed convex hull* of the *level set* $L(x_0)$. The problem of minimizing f over E can of course be reduced to that over $W(x_0)$, which we shall often assume is bounded. We shall always assume that f is bounded below, so that we can speak of trying to minimize f. The goal of our analysis of each method will be to compute a sequence $\{x_n\}$ such that $\{f(x_n)\}$ is decreasing, hopefully toward the infimum of f. Generally we shall discover that, for some $\delta > 0$, we have

$$f(x_n) - f(x_{n+1}) \geq -\delta\langle p_n(x_n), \nabla f(x_n)\rangle \| p_n(x_n) \|^{-1}$$

If f is bounded below, then $f(x_n) - f(x_{n+1})$ and hence $\langle p_n(x_n), \nabla f(x_n)\rangle \| p_n(x_n) \|^{-1}$ must converge to zero; from the admissibility of $\{p_n(x_n)\}$ we can then conclude that $\| \nabla f(x_n) \| \longrightarrow 0$. If $f(x_n) \longrightarrow \inf_{x \in E} f(x)$, we call $\{x_n\}$ a minimizing sequence; if $\nabla f(x) = 0$, x is called a *critical point* [Vainberg (1964)]. Thus we are led to the following definition.

DEFINITION 4.2.1. A sequence $\{x_n\}$ is called a *criticizing sequence* if and only if $\| \nabla f(x_n) \| \longrightarrow 0$.

Thus our numerical methods to be discussed will provide us merely with criticizing sequences; we wish to know under what circumstances this yields indeed a minimizing sequence.

THEOREM 4.2.1. If $W(x_0)$ is bounded and $\{x_n\}$ is a criticizing sequence for the convex functional f bounded below, then $\{x_n\}$ is a minimizing sequence.

Proof: Let $y_n \in W(x_0)$ be chosen such that $\lim_{n\to\infty} f(y_n) = \inf_{x \in E} f(x)$. By the convexity of f in E we have from Proposition 1.5.1 that

$$f(y_n) - f(x_n) \geq \langle y_n - x_n, \nabla f(x_n) \rangle$$

Since $\| y_n - x_n \|$ is bounded and $\| \nabla f(x_n) \| \to 0$, we have

$$\inf_{x \in E} f(x) \leq \liminf_{n\to\infty} f(x_n) \leq \limsup_{n\to\infty} f(x_n) \leq \lim_{n\to\infty} f(y_n) = \inf_{x \in E} f(x) \quad \text{Q.E.D.}$$

THEOREM 4.2.2. If E is reflexive, if $\nabla f(x) = 0$ at only one point in $W(x_0)$, if $W(x_0)$ is bounded, and if $x_n \rightharpoonup x^*$ implies $\langle y, \nabla f(x_n) \rangle \to \langle y, \nabla f(x^*) \rangle$ for all $y \in E$, then each criticizing sequence converges weakly to the unique point x^* minimizing f over $W(x_0)$.

Proof: Since $W(x_0)$ is a convex, bounded, norm-closed subset of reflexive space, it is weakly compact and thus some subsequence $x_{n_i} \rightharpoonup x' \in W(x_0)$. Then since $\| \nabla f(x_n) \| \to 0$, for each y in E we have

$$0 = \lim_{n_i \to \infty} \langle y, \nabla f(x_{n_i}) \rangle = \langle y, \nabla f(x') \rangle$$

and hence $\nabla f(x') = 0$. This then implies $x' = x^*$ for every weak limit point x' of $\{x_n\}$, and hence $x_n \rightharpoonup x^*$. Q.E.D.

Further results on the convergence of criticizing sequences can be obtained by considering properties of ∇f, as in Theorem 4.2.2, or by using Theorem 4.2.1 in conjunction with statements about minimizing sequences in Section 1.6. Therefore, in what follows we shall often go no further than making statements about criticizing sequences.

EXERCISE. Derive further results concerning the convergence of criticizing sequences by considering properties of ∇f, as in Theorem 4.2.2, and by using Theorem 4.2.1 together with results in Section 1.6.

We shall often prove that certain criticizing sequences satisfy

$$\lim_{n\to\infty} \| x_{n+1} - x_n \| = 0$$

a fact which is useful in many cases since such a sequence cannot have two distinct norm-limit points x' and x'' unless there is a continuum of limit points "connecting" x' and x''. For, if there are only finitely many limit points $x^{(1)}, \ldots, x^{(N)}$, then there is an $\epsilon > 0$ with $\|x^{(i)} - x^{(j)}\| \geq \epsilon$ if $i \neq j$; for large enough n, x_n must lie in some one of the spheres of radius $\epsilon/3$ about the $x^{(i)}$, $i = 1, 2, \ldots, N$. Since $\|x_{n+1} - x_n\| \longrightarrow 0$, this implies in fact that all the x_n must be in some one fixed sphere, since to jump to another one requires $\|x_{n+1} - x_n\| \geq \epsilon/3$ which is never true for large n. Although we shall improve this result later, we have proved the following theorem.

THEOREM 4.2.3. If $\nabla f(x)$ is norm-continuous in x, if $\nabla f(x) = 0$ has only finitely many solutions in $W(x_0)$, and if $\{x_n\}$ is a criticizing sequence with $\|x_{n+1} - x_n\| \to 0$, then $\{x_n\}$ either has no norm-limit points or $x_n \longrightarrow x^*$ with $\nabla f(x^*) = 0$.

We do not wish to give the impression that the only way to treat minimization is from the criticizing-sequence point of view; other approaches also can be taken. For example [Yakovlev (1965)], suppose the directions p_n are generated via $p_n = -H_n \nabla f(x_n)$ where for each n, H_n is a bounded, positive-definite self-adjoint linear operator from the Hilbert space E into itself. Suppose

$$0 < a\langle H_n^{-1}x, x\rangle \leq \langle f_y''x, x\rangle \leq A\langle H_n^{-1}x, x\rangle$$

for all x, y in E, and suppose we take

$$x_{n+1} = x_n + t_n p_n, \qquad 0 < \epsilon_1 \leq t_n \leq \frac{2}{A} - \epsilon_2$$

Then f of course is uniquely minimized at a point x^* and one can prove that $x_n \longrightarrow x^*$. Arguing much as we shall in Section 4.6, one can show that

$$f(x_{n+1}) - f(x_n) \leq -t_n\left(1 - \frac{t_n A}{2}\right)\langle H_n \nabla f(x_n), \nabla f(x_n)\rangle$$

and

$$\langle H_n \nabla f(x_n), \nabla f(x_n)\rangle \geq a^2 \langle H_n^{-1}(x_n - x^*), x_n - x^*\rangle$$

Since $\nabla f(x^*) = 0$, we have

$$\begin{aligned} f(x_n) - f(x^*) &\leq \frac{A}{2}\langle H_n^{-1}(x_n - x^*), x_n - x^*\rangle \\ &\leq \frac{A}{2a^2}\langle H_n \nabla f(x_n), \nabla f(x_n)\rangle \end{aligned}$$

Therefore,

$$f(x_{n+1}) - f(x_n) \leq -t_n\left(1 - \frac{t_n A}{2}\right)\frac{2a^2}{A}[f(x_n) - f(x^*)]$$

which yields

$$f(x_{n+1}) - f(x^*) \leq q[f(x_n) - f(x^*)]$$

for a certain $q < 1$, since $0 < \epsilon_1 \leq t_n \leq (2/A) - \epsilon_2$. Thus

$$f(x_n) - f(x^*) \leq q^n[f(x_0) - f(x^*)]$$

and the sequence is a minimizing sequence. Since

$$f(x_n) - f(x^*) \geq \frac{a}{2}\langle H_i^{-1}(x_n - x^*), x_n - x^*\rangle$$

it follows that $x_n \longrightarrow x^*$. Thus convergence proofs can be given by estimating $f(x_n) - f(x^*)$ directly rather than using the criticizing-sequence approach. It is true, however, that the direction sequence so generated is admissible, implying that the analysis is possible from either viewpoint.

Historically, most convergence proofs for the step-size algorithms we are about to consider have been performed by contradiction; this is rigorous but often not intuitive. Recently it has been shown [Céa (1969)] that a single unifying principle can be used to analyze directly many methods; we shall use an extended version of this approach whenever possible.

DEFINITION 4.2.2. A function $c(t)$, defined for $t \geq 0$, is called a *forcing function* if and only if $c(t) \geq 0$, and $c(t_n)$ can converge to zero only if t_n converges to zero.

EXERCISE. Give some examples of forcing functions.

Throughout this chapter we shall be assuming that ∇f is uniformly continuous on $W(x_0)$; this implies that for every $\epsilon > 0$ there exists $\delta > 0$ such that $||x - y|| < \delta$ implies $||\nabla f(x) - \nabla f(y)|| < \epsilon$ for $x, y \in W(x_0)$. In particular, we can let $\delta = s(\epsilon)$, where $s(t)$ is the forcing function (*reverse modulus of continuity*) defined by the following.

DEFINITION 4.2.3.

$$s(t) \equiv \inf\{||x - y||; x, y \in W(x_0), ||\nabla f(x) - \nabla f(y)|| \geq t\}$$

EXERCISE. Prove that $s(t)$ is a monotone nonincreasing forcing function and that we can set $\delta = s(\epsilon)$ in the description above of uniform continuity of ∇f.

In terms of these concepts, we can prove the following theorem, which will be our fundamental tool in subsequent sections.

THEOREM 4.2.4. Let f be bounded below on $W(x_0)$, let ∇f be uniformly continuous on $W(x_0)$, and let $p_n = p_n(x_n)$ be an admissible direction sequence. Let there exist functions $c_1(t)$ and $c_2(t)$ such that $c_1(t)$ and $t - c_2(t)$ are forcing functions. (A) If the step sizes satisfy

$$c_1\left(\left\langle -\nabla f(x_n), \frac{p_n}{\|p_n\|}\right\rangle\right) \leq t_n \|p_n\| \leq s\left[c_2\left(\left\langle -\nabla f(x_n), \frac{p_n}{\|p_n\|}\right\rangle\right)\right]$$

then $\{x_n\}$ is a criticizing sequence. (B) If t'_n is any step size so that

$$c_1\left(\left\langle -\nabla f(x_n), \frac{p_n}{\|p_n\|}\right\rangle\right) \leq t'_n \|p_n\| \leq s\left[c_2\left(\left\langle -\nabla f(x_n), \frac{p_n}{\|p_n\|}\right\rangle\right)\right]$$

and if x_{n+1} is chosen as any point such that

$$f(x_n) - f(x_{n+1}) \geq \beta[f(x_n) - f(x_n + t'_n p_n)] \quad \text{for} \quad \beta > 0$$

then $\{x_n\}$ is a criticizing sequence. In either case A or B,

$$f(x_n) - f(x_{n+1}) \geq \lambda t_n \|p_n\| [\gamma_n - c_2(\gamma_n)] \geq \lambda c_1(\gamma_n)[\gamma_n - c_2(\gamma_n)]$$

where

$$\gamma_n \equiv \left\langle -\nabla f(x_n), \frac{p_n}{\|p_n\|}\right\rangle$$

and $\lambda = 1$ or β in cases A or B, respectively.

Proof: First we consider case A. Since $t_n \|p_n\| \leq s[c_2(\gamma_n)]$, we have

$$\left|\left\langle \nabla f(x_n + tp_n) - \nabla f(x_n), \frac{p_n}{\|p_n\|}\right\rangle\right| \leq c_2(\gamma_n) \quad \text{for} \quad 0 \leq t \leq t_n$$

and hence

$$\left\langle -\nabla f(x_n + tp_n), \frac{p_n}{\|p_n\|}\right\rangle \geq \gamma_n - c_2(\gamma_n) \quad \text{for} \quad 0 \leq t \leq t_n$$

Since $f(x_n) - f(x_{n+1}) = \langle -\nabla f(x_n + tp_n), t_n p_n\rangle$ for *some* t in $(0, t_n)$ and $t_n \|p_n\| \geq c_1(\gamma_n)$, we have

$$f(x_n) - f(x_{n+1}) \geq t_n \|p_n\| [\gamma_n - c_2(\gamma_n)] \geq c_1(\gamma_n)[\gamma_n - c_2(\gamma_n)] \geq 0$$

Hence $\gamma_n \longrightarrow 0$ and $\{x_n\}$ is criticizing. Now consider case B. We can, by case A, write

$$f(x_n) - f(x_{n+1}) \geq \beta c_1(\gamma_n)[\gamma_n - c_2(\gamma_n)]$$

and we are again done. Q.E.D.

Remark 1. We point out that we can get by with weaker assumptions than the admissibility of the direction sequence. In Theorem 4.2.4 we found that

$$f(x_n) - f(x_{n+1}) \geq c(\gamma_n)$$

for some forcing function $c(t)$. This implies that

$$\sum_{n=0}^{\infty} c(\gamma_n) \leq \sum_{n=0}^{\infty} [f(x_n) - f(x_{n-1})] < \infty$$

and thus we only need choose directions such that $\sum_{n=0}^{\infty} c(\gamma)_n < \infty$ implies $\| \nabla f(x_n) \| \longrightarrow 0$. In particular, in the theorems in the next sections, we shall find $c(t) = \text{const} \cdot ts(t)$, where $s(t)$ is the reverse modulus of continuity of ∇f; if ∇f is Lipschitz-continuous, we have $s(t) = \text{const} \cdot t$ and hence $c(t) = \text{const} \cdot t^2$. Therefore, if

$$\sum_{n=0}^{\infty} \alpha_n^2 = \infty \quad \text{where} \quad \alpha_n = \frac{\langle \nabla f(x_n), p_n \rangle}{\| \nabla f(x_n) \| \| p_n \|}$$

then $\sum_{n=0}^{\infty} c(\gamma_n) < \infty$ implies $\sum_{n=0}^{\infty} \alpha_n^2 \| \nabla f(x_n) \|^2 < \infty$ which implies $\| \nabla f(x_n) \| \longrightarrow 0$. Generally speaking, however, we do not feel that this analysis is applicable to many direction algorithms; in our experience, direction sequences used in practice are usually admissible in our sense. We shall not, therefore, state theorems based on the fact that $\sum_{n=0}^{\infty} c(\gamma_n) < \infty$; the reader should, however, be aware of this approach.

Remark 2. In the following sections we shall be proving that various choices of t_n yield criticizing sequences. By part B of Theorem 4.2.4, there is always the obvious corollary concerning the choice of x_{n+1}; although this is a useful fact since, in particular, it indicates that t_n need not be found exactly, we shall not bore the reader by continually stating this corollary. It should, however, be remembered.

4.3. GLOBAL MINIMUM ALONG THE LINE

We consider first intuitively the most natural way of choosing t_n, by minimizing $f(x_n + tp_n)$ as a function of $t \geq 0$; we assume that such t_n always exists. We prove a more general theorem.

EXERCISE. Prove that t_n exists if $W(x_0)$ is bounded.

THEOREM 4.3.1. Let f be bounded below on $W(x_0)$, let ∇f be uniformly continuous on $W(x_0)$, and let $p_n = p_n(x_n)$ define an admissible sequence of directions. For a set of numbers $\alpha_n \in [0, \alpha]$ with $\alpha < 1$, choose t_n so that

$$f(x_n + t_n p_n) - \alpha_n t_n \langle \nabla f(x_n), p_n \rangle \leq f(x_n + tp_n) - \alpha_n t \langle \nabla f(x_n), p_n \rangle$$

for all $t \geq 0$. Then $\{x_n\}$ is a criticizing sequence and

$$f(x_n) - f(x_{n+1}) \geq s(c\gamma_n)(1 - c)\gamma_n$$

for all c in $(0, 1 - \alpha)$, where we let

$$\gamma_n \equiv \left\langle -\nabla f(x_n), \frac{p_n}{\|p_n\|} \right\rangle$$

Proof: Since t_n minimizes $f(x_n + tp_n) - \alpha_n t \langle \nabla f(x_n), p_n \rangle$, we have

$$\langle \nabla f(x_n + t_n p_n), p_n \rangle - \alpha_n \langle \nabla f(x_n), p_n \rangle = 0$$

For any fixed c in $(0, 1 - \alpha)$, if $t_n \|p_n\| < s(c\gamma_n)$, we would have

$$\left| \left\langle [\nabla f(x_n + t_n p_n) - \alpha_n \nabla f(x_n)] - [\nabla f(x_n) - \alpha_n \nabla f(x_n)], \frac{p_n}{\|p_n\|} \right\rangle \right| \leq c\gamma_n$$

This would then give $(1 - \alpha)\gamma_n \leq (1 - \alpha_n)\gamma_n \leq c\gamma_n$, a contradiction to $c \in (0, 1 - \alpha)$. Therefore, $t_n \|p_n\| \geq s(c\gamma_n)$. By part A of Theorem 4.2.4 with $c_1(t) \equiv s(ct)$ and $c_2(t) \equiv ct$, the method generated by $t'_n \|p_n\| = s(c\gamma_n)$ yields a criticizing sequence. By the defining property of t_n, we have

$$\begin{aligned} f(x_n) - f(x_n + t_n p_n) &\geq f(x_n) - f(x_n + t'_n p_n) + \alpha_n(t_n - t'_n)\langle -\nabla f(x_n), p_n \rangle \\ &\geq f(x_n) - f(x_n + t'_n p_n) \end{aligned}$$

since we showed above that $t_n \geq t'_n$. The theorem now follows from part B of Theorem 4.2.4 with $\beta \equiv 1$. Q.E.D.

EXERCISE. Assuming ∇f to be Lipschitzian, apply the approach of Remark 1 after Theorem 4.2.4 to derive another convergence theorem for the method of Theorem 4.3.1.

Remark. Setting $\alpha_n \equiv \alpha = 0$ yields the usual method.

General references: Altman (1966a), Céa (1969), Elkin (1968), Goldstein (1964b, 1965, 1966, 1967).

4.4. FIRST LOCAL MINIMUM ALONG THE LINE: POSITIVE WEIGHT THEREON

The problem of locating the absolute minimum along $x_n + tp_n$ is quite difficult unless one knows, for example, that every local minimum is a global minimum. In any case, it would be simpler to seek the first local minimum—

say, by using a one-dimensional root-finding method to locate the first root of $\langle \nabla f(x_n + tp_n, p_n \rangle = 0$; since such a root may give only an inflection point, we must analyze this possibility also.

EXERCISE. Find conditions under which the first *local* minimum with respect to t along $x_n + tp_n$ is a *global* one.

Actually, as we saw in Remark 2 following Theorem 4.2.4, it is not vital to reach the local minimum exactly. We analyze some additional ways to describe how close one need come.

THEOREM 4.4.1. Let f be bounded below on $W(x_0)$, let ∇f be uniformly continuous on $W(x_0)$, and let $p_n = p_n(x_n)$ define an admissible direction sequence. Let t_n be either (1) the smallest positive t_n providing a local minimum for

$$f(x_n + tp_n) - \alpha_n t \langle \nabla f(x_n), p_n \rangle$$

or (2) the first positive root of

$$\langle \nabla f(x_n + tp_n), p_n \rangle - \alpha_n \langle \nabla f(x_n), p_n \rangle = 0$$

or (3) the following:

$$t_n = \sup \{t; \langle \nabla f(x_n + \tau p_n), p_n \rangle - \alpha_n \langle \nabla f(x_n), p_n \rangle \leq 0 \text{ for } 0 \leq \tau \leq t\}$$

We assume $0 \leq \alpha_n \leq \alpha < 1$. Let $x_{n+1} = x_n + t_n p_n$; then $\{x_n\}$ is a criticizing sequence and

$$f(x_n) - f(x_{n+1}) \geq s(c\gamma_n)(1 - c)\gamma_n$$

for all $c \in (0, 1 - \alpha)$ with

$$\gamma_n \equiv \left\langle -\nabla f(x_n), \frac{p_n}{||p_n||} \right\rangle$$

Proof: In any determination of t_n, clearly

$$\langle \nabla f(x_n + t_n p_n), p_n \rangle - \alpha_n \langle \nabla f(x_n), p_n \rangle = 0$$

and

$$\frac{d}{dt}[f(x_n + tp_n) - \alpha_n t \langle \nabla f(x_n), p_n \rangle] \leq 0$$

for $0 \leq t \leq t_n$, implying

$$f(x_n + t_n p_n) - \alpha_n t_n \langle \nabla f(x_n), p_n \rangle \leq f(x_n + tp_n) - \alpha_n t \langle \nabla f(x_n), p_n \rangle$$

for $0 \leq t \leq t_n$. The proof is now identical to that of Theorem 4.3.1. Q.E.D.

EXERCISE. As a generalization of Theorems 4.3.1 and 4.4.1, show that t_n may be chosen as any number satisfying

$$\langle \nabla f(x_n + t_n p_n\rangle - \alpha_n \langle \nabla f(x_n), p_n\rangle = 0$$

and

$$f(x_n + t_n p_n) - \alpha_n t_n \langle \nabla f(x_n), p_n\rangle \leq f(x_n + tp_n) - \alpha_n t \langle \nabla f(x_n), p_n\rangle$$

for $0 \leq t \leq t_n$, where $\alpha_n \in [0, \alpha]$ and $\alpha < 1$.

EXERCISE. Assuming ∇f to be Lipschitzian, apply the approach of Remark 1 after Theorem 4.2.4 to find another convergence theorem for the method of Theorem 4.4.1.

As a special case, one may take $\alpha_n \equiv \alpha \in [0, 1)$, a method often called the *generalized Curry method* in recognition of Curry's result [Curry (1944)] with $\alpha = 0$ in which one essentially seeks the first local minimum of $f(x_n + tp_n)$. We state this as a corollary.

COROLLARY 4.4.1. Let f be bounded below on $W(x_0)$, let ∇f be uniformly continuous on $W(x_0)$, and let $p_n = p_n(x_n)$ define an admissible direction sequence. Let t_n be defined as either (1) the smallest positive t_n providing a local minimum for $f(x_n + tp_n)$; or (2) the first positive root of

$$\langle \nabla f(x_n + tp_n), p_n\rangle = 0$$

or (3) the following:

$$t_n = \sup \{t; \langle \nabla f(x_n + \tau p_n), p_n\rangle \leq 0 \quad \text{for} \quad 0 \leq \tau \leq t\}$$

Let $x_{n+1} = x_n + t_n p_n$; then $\{x_n\}$ is a criticizing sequence and

$$f(x_n) - f(x_{n+1}) \geq s(c\gamma_n)(1 - c)\gamma_n$$

for all c in (0, 1) with

$$\gamma_n \equiv \left\langle -\nabla f(x_n), \frac{p_n}{\|p_n\|}\right\rangle$$

Another method of describing the weight on the minimum is as follows. First, choose a t'_n precisely as t_n is determined in the Theorem 4.4.1, above, with $0 \leq \alpha_n \leq \alpha < 1$; then define $t_n = \lambda_n t'_n$ for an appropriate *relaxation factor* λ_n.

THEOREM 4.4.2. Under the hypotheses of the preceding theorem, let t'_n be determined as is t_n there. Let $t_n = \lambda_n t'_n$ where $d(\gamma_n) \leq \lambda_n \leq 1$ for some forcing function $d(t)$ with

$$\gamma_n \equiv \left\langle -\nabla f(x_n), \frac{p_n}{\|p_n\|} \right\rangle$$

Let $x_{n+1} = x_n + t_n p_n$; then $\{x_n\}$ is a criticizing sequence and

$$f(x_n) - f(x_{n+1}) \geq d(\gamma_n)s(c\gamma_n)(1-c)\gamma_n$$

for all c in $(0, 1-\alpha)$.

Proof: From the proof of the preceding theorem we know that $t'_n \|p_n\| \geq s(c\gamma_n)$ for all c in $(0, 1-\alpha)$; therefore,

$$t'_n \|p_n\| \geq t_n \|p_n\| \geq d(\gamma_n)s(c\gamma_n)$$

Therefore, $t''_n \|p_n\| \equiv d(\gamma_n)s(c\gamma_n)$ yields a criticizing sequence by part A of Theorem 4.2.4 with $c_1(t) \equiv d(t)s(ct)$ and $c_2(t) \equiv ct$. Since

$$\frac{d}{dt}[f(x_n + tp_n) - \alpha_n t\langle \nabla f(x_n), p_n \rangle] \leq 0 \quad \text{for} \quad 0 \leq t \leq t_n$$

then

$$f(x_{n+1}) - \alpha_n t_n \langle \nabla f(x_n), p_n \rangle \leq f(x_n + t''_n p_n) - \alpha_n t''_n \langle \nabla f(x_n), p_n \rangle$$

and thus

$$f(x_n) - f(x_{n+1}) \geq f(x_n) - f(x_n + t''_n p_n)$$

yielding convergence by part B of Theorem 4.2.4 with $\beta = 1$. Q.E.D.

General references: Altman (1966a), Elkin (1968), Goldstein (1964b, 1965, 1966, 1967), Levitin–Poljak (1966a).

4.5. A SIMPLE INTERVAL ALONG THE LINE

In some cases it is possible to write down beforehand a simple interval from which t_n can be chosen arbitrarily, guaranteeing the generation of a criticizing sequence. If ∇f satisfies $\|\nabla f(x) - \nabla f(y)\| \leq L\|x - y\|$ in $W(x_0)$ and L is the best such constant, then we have from Definition 4.2.3 that $s(t) \geq t/L$. Our Theorem 4.2.4 then tells us that the choice of t_n so that, for example,

$$\epsilon_1 \gamma_n \leq t_n \|p_n\| \leq \frac{1-\epsilon_2}{L}\gamma_n, \quad \text{with} \quad \epsilon_1 > 0, \quad \epsilon_2 > 0$$

and

$$\gamma_n \equiv \left\langle -\nabla f(x_n), \frac{p_n}{\|p_n\|} \right\rangle$$

will yield a criticizing sequence. By more careful analysis, the size of this interval can be doubled, as we now proceed to show.

THEOREM 4.5.1. Let f be bounded below on $W(x_0)$, let ∇f satisfy

$$\|\nabla f(x) - \nabla f(y)\| \leq L\|x - y\| \quad \text{for} \quad x, y \in W(x_0)$$

and suppose the directions $p_n(x)$ satisfy

$$\|p_n(x_n)\| \leq A_1 \|\nabla f(x_n)\|, \qquad -\langle \nabla f(x_n), p_n(x_n)\rangle \geq A_2 \|\nabla f(x_n)\|^2, \qquad A_2 > 0$$

Then if $\delta_1 > 0$ and $\delta_2 > 0$ and if t_n is chosen with

$$\delta_1 \leq t_n \leq \frac{2A_2}{LA_1^2} - \delta_2$$

then the sequence $\{x_n\}$ is criticizing, $\|x_{n+1} - x_n\| \longrightarrow 0$, and

$$f(x_n) - f(x_{n+1}) \geq \epsilon \|\nabla f(x_n)\|^2 \quad \text{for some} \quad \epsilon > 0$$

Proof:

$$\begin{aligned} f(x_{n+1}) &= f(x_n) + \int_0^1 \langle \nabla f(x_n + \alpha t_n p_n), t_n p_n \rangle \, d\alpha \\ &= f(x_n) + \langle \nabla f(x_n), t_n p_n \rangle + \int_0^1 \langle \nabla f(x_n + \alpha t_n p_n) - \nabla f(x_n), t_n p_n \rangle \, d\alpha \\ &\leq f(x_n) - t_n A_2 \|\nabla f(x_n)\|^2 + L t_n^2 \|p_n\|^2 \int_0^1 \alpha \, d\alpha \\ &\leq f(x_n) - t_n \|\nabla f(x_n)\|^2 \left[A_2 - t_n \frac{L}{2} A_1^2 \right] \end{aligned}$$

For t_n in the given interval, the term in brackets is bounded away from zero below, so $f(x_{n+1}) < f(x_n)$ and $\|\nabla f(x_n)\| \longrightarrow 0$. Since

$$\|p_n(x_n)\| \leq A_1 \|\nabla f(x_n)\| \longrightarrow 0$$

and $|t_n|$ is bounded,

$$\|x_{n+1} - x_n\| = \|t_n p_n(x_n)\| \longrightarrow 0$$

Q.E.D.

EXERCISE. Apply the approach of Remark 1 after Theorem 4.2.4 to find another convergence theorem for the method of Theorem 4.5.1.

A similar simple range is given as follows.

THEOREM 4.5.2. Let f be bounded below on $W(x_0)$, let ∇f satisfy $\|\nabla f(x) - \nabla f(y)\| \leq L\|x - y\|$, and let

$$x_{n+1} = x_n - T_n\langle \nabla f(x_n), p_n\rangle p_n$$

where $p_n \equiv p_n(x_n)$ is an admissible sequence of directions with $\|p_n\| = 1$ and

$$0 < \delta_1 \leq T_n \leq \frac{2}{L} - \delta_2 < \frac{2}{L}$$

The $\{x_n\}$ is a criticizing sequence, $\|x_{n+1} - x_n\| \longrightarrow 0$, and

$$f(x_n) - f(x_{n+1}) \geq \epsilon\|\nabla f(x_n)\|^2 \quad \text{for some} \quad \epsilon > 0$$

Proof: Proceeding just as above via integration we find

$$\begin{aligned} f(x_{n+1}) &\leq f(x_n) - T_n\langle \nabla f(x_n), p_n\rangle^2 + \tfrac{1}{2}LT_n^2\langle \nabla f(x_n), p_n\rangle^2 \\ &\leq f(x_n) - T_n\langle \nabla f(x_n), p_n\rangle^2[1 - \tfrac{1}{2}LT_n] \end{aligned}$$

Thus $f(x_{n+1}) < f(x_n)$, and $\langle \nabla f(x_n), p_n\rangle$ and $\|\nabla f(x_n)\|$ converge to zero. Finally,

$$\|x_{n+1} - x_n\| = \|T_n\langle \nabla f(x_n), p_n\rangle p_n\| \leq \frac{2}{L}|\langle \nabla f(x_n), p_n\rangle| \longrightarrow 0$$

Q.E.D.

Remark. It is a simple matter to allow a slightly larger range for T_n in Theorem 4.5.2—namely,

$$s(\langle -\nabla f(x_n), p_n\rangle) \leq T_n \leq \frac{2}{L} - s(\langle -\nabla f(x_n), p_n\rangle)$$

where $0 \leq \dot{s}(t) \leq 1$, and $s(t)$ is a forcing function [Elkin (1968)].

EXERCISE. Prove the assertion in the above *Remark*.

Finally, we state one more result giving a simple range but not depending on a priori knowledge of Lipschitz constants.

THEOREM 4.5.3. Suppose f is a convex functional bounded below on $W(x_0)$ and such that $\nabla f(x)$ is uniformly continuous in x in $W(x_0)$. Let $p_n = p_n(x_n)$ be an admissible sequence of directions with $\|p_n\| = 1$, and pick δ_1, δ_2 satisfying $0 < \delta_1 < \delta_2 < 1$. Let $x_{n+1} = x_n + T_n p_n$ where T_n is determined by:

1. if $-\langle \nabla f(x_n + \langle -\nabla f(x_n), p_n \rangle p_n), p_n \rangle \geq \delta_1 \langle -\nabla f(x_n), p_n \rangle$, then set $T_n = \langle -\nabla f(x_n), p_n \rangle$;
2. otherwise compute T_n, which always exists, to satisfy $0 < T_n < -\langle \nabla f(x_n), p_n \rangle$

and

$$-\delta_1 \langle \nabla f(x_n), p_n \rangle \leq -\langle \nabla f(x_n + T_n p_n), p_n \rangle \leq -\delta_2 \langle \nabla f(x_n), p_n \rangle$$

Then $\{x_n\}$ is a criticizing sequence and $\|x_{n+1} - x_n\| \longrightarrow 0$.

Proof: By Proposition 1.5.1 we have

$$\langle \nabla f(x_n + t'' p_n) - \nabla f(x_n + t' p_n), p_n \rangle \geq 0 \quad \text{if} \quad t'' \geq t'$$

Therefore,

$$\begin{aligned} f(x_n) - f(x_{n+1}) &= \langle \nabla f(x_n + \lambda T_n p_n), -T_n p_n \rangle \\ &\geq -T_n \langle \nabla f(x_n + T_n p_n), p_n \rangle \\ &\geq T_n \delta_1 \langle -\nabla f(x_n), p_n \rangle \end{aligned}$$

Thus $\{f(x_n)\}$ is decreasing and hence convergent and $T_n \delta_1 \langle -\nabla f(x_n), p_n \rangle$ tends to zero. If infinitely often we have

$$-\langle \nabla f(x_n), p_n \rangle \geq \epsilon > 0$$

then it cannot occur under condition 1 since then we have

$$f(x_n) - f(x_{n+1}) \geq \delta_1 \langle \nabla f(x_n), p_n \rangle^2 \geq \delta_1 \epsilon^2$$

in contradiction to the boundedness below of f. Under condition 2, however, we have

$$\epsilon(1 - \delta_2) \leq (1 - \delta_2) \langle -\nabla f(x_n), p_n \rangle \leq \langle \nabla f(x_n + T_n p_n) - \nabla f(x_n), p_n \rangle$$

the right-hand side of which tends to zero since ∇f is uniformly continuous and since T_n must tend to zero if $\langle -\nabla f(x_n), p_n \rangle$ does not. This gives a contradiction, leading us to conclude that $\langle -\nabla f(x_n), p_n \rangle$ and hence $\|\nabla f(x_n)\|$ tends to zero. Finally,

$$\|x_{n+1} - x_n\| = \|T_n p_n\| = |T_n| \leq |\langle \nabla f(x_n), p_n \rangle| \longrightarrow 0$$

Q.E.D.

General references: Altman (1966a), Elkin (1968), Goldstein (1964b, 1965, 1966, 1967), Levitin–Poljak (1966a).

4.6. A RANGE FUNCTION ALONG THE LINE

We shall now describe another way of selecting t_n by making use of a function $g(x, t, p)$ which will determine the range of values t can assume. The method is similar to that in Theorem 4.5.3 except that a different measure of the distance to be moved is used. The main idea is to pick t_n to guarantee that the decrease in f dominates

$$\left\langle \nabla f(x_n), \frac{p_n}{\| p_n \|} \right\rangle$$

as discussed in Section 4.2.

We shall determine admissible values of t_n in terms of the *range function*

$$g(x, t, p) \equiv \frac{f(x) - f(x + tp)}{-t\langle \nabla f(x), p \rangle}$$

which is continuous at $t = 0$ if we define $g(x, 0, p) = 1$. We shall assume that an admissible sequence of directions p_n is given satisfying $\| p_n \| = 1$. Given a number δ satisfying $0 < \delta \leq \frac{1}{2}$ and a forcing function d satisfying $d(t) \leq \delta t$, we shall attempt to move from x_n to $x_{n+1} = x_n + t_n p_n$ as follows: if, for $t_n = 1$ and $x'_n \equiv x_n + p_n$ we find

$$g(x_n, t_n, p_n) \geq \frac{d(\langle -\nabla f(x_n), p_n \rangle)}{\langle -\nabla f(x_n), p_n \rangle} \tag{4.6.1}$$

then we set $x_{n+1} = x'_n$; otherwise, find $t_n \in (0, 1)$ satisfying Equation 4.6.1 and also

$$| g(x_n, t_n, p_n) - 1 | \geq \frac{d(\langle -\nabla f(x_n), p_n \rangle)}{\langle -\nabla f(x_n), p_n \rangle} \tag{4.6.2}$$

First we observe that this algorithm is well defined. Since $g(x_n, 0, p_n) = 1$, and

$$1 - \frac{d(t)}{t} \geq \frac{d(t)}{t} \qquad \text{for all } t$$

if we have

$$g(x_n, 1, p_n) < \frac{d(z)}{z}$$

where $z = \langle -\nabla f(x_n), p_n \rangle$, then by the continuity of $g(x_n, t, p_n)$ in t there exists $t_n \in (0, 1)$ with

$$\frac{d(z)}{z} \leq g(x_n, t_n, p_n) \leq 1 - \frac{d(z)}{z}$$

Now we prove the convergence of the method.

THEOREM 4.6.1 [Elkin (1968)]. Let f be bounded below on $W(x_0)$, $\nabla f(x)$ be uniformly continuous in x in $W(x_0)$, and $p_n = p_n(x_n)$ give an admissible sequence of directions with $\|p_n\| = 1$. Let $d(t)$ be a forcing function with $d(t) \leq \delta t$, $0 < \delta \leq \frac{1}{2}$. If, for $t_n = 1$, Equation 4.6.1 is valid, let $x_{n+1} = x'_n \equiv x_n + p_n$. Otherwise, find $t_n \in (0, 1)$ satisfying Equation 4.6.1 and Equation 4.6.2. Then $\{x_n\}$ is a criticizing sequence, and

$$f(x_n) - f(x_{n+1}) \geq \lambda_n d(\langle -\nabla f(x_n), p_n \rangle)$$

where $\lambda_n = 1$ if $t_n = 1$ and $\lambda_n = s(d(\langle -\nabla f(x_n), p_n \rangle))$ if $t_n \neq 1$, where s is the reverse modulus of continuity of ∇f.

Proof: By Equation 4.6.1, $f(x_n)$ is decreasing and

$$f(x_n) - f(x_{n+1}) \geq t_n d(\langle -\nabla f(x_n), p_n \rangle) \tag{4.6.3}$$

If $t_n = 1$ does not satisfy Equation 4.6.1, then $t_n \in (0, 1)$. For these n, we write

$$f(x_{n+1}) - f(x_n) = \langle \nabla f(x_n + \lambda_n t_n p_n), t_n p_n \rangle \quad \text{for some} \quad \lambda_n \in (0, 1)$$

Thus, from Equation 4.6.2,

$$\begin{aligned} \frac{d(\langle -\nabla f(x_n), p_n \rangle)}{\langle -\nabla f(x_n), p_n \rangle} &\leq | g(x_n, t_n, p_n) - 1 | \\ &\leq \left| \frac{\langle \nabla f(x_n + \lambda_n t_n p_n) - \nabla f(x_n), p_n \rangle}{\langle \nabla f(x_n), p_n \rangle} \right| \\ &\leq \frac{\| \nabla f(x_n + \lambda_n t_n p_n) - \nabla f(x_n) \|}{\langle -\nabla f(x_n), p_n \rangle} \end{aligned}$$

Therefore,

$$\| \nabla f(x_n + \lambda_n t_n p_n) - \nabla f(x_n) \| \geq d(\langle -\nabla f(x_n), p_n \rangle)$$

and hence

$$\begin{aligned} t_n = \| x_{n+1} - x_n \| \geq \| \lambda_n t_n p_n \| &\geq s[\| \nabla f(x_n + \lambda_n t_n p_n) - \nabla f(x_n) \|] \\ &\geq s[d(\langle -\nabla f(x_n), p_n \rangle)] \end{aligned} \tag{4.6.4}$$

Hence, using Equation 4.6.3, we conclude that

$$f(x_n) - f(x_{n+1}) \geq d(\langle -\nabla f(x_n), p_n \rangle) s[d(\langle -\nabla f(x_n), p_n \rangle)]$$

Thus

$$f(x_n) - f(x_{n+1}) \geq \lambda_n d(\langle -\nabla f(x_n), p_n \rangle)$$

as asserted, which implies, as before, $\langle -\nabla f(x_n), p_n \rangle \to 0$. Q.E.D.

Computationally one needs a procedure for computing a $t_n \in (0, 1)$ satisfying Equations 4.6.1 and 4.6.2 if $t_n = 1$ does not satisfy Equation 4.6.1. We consider doing this [Armijo (1966), Elkin (1968)] by successively trying the values

$$t_n = \alpha, \alpha^2, \alpha^3, \ldots, \quad \text{for some} \quad \alpha \in (0, 1)$$

THEOREM 4.6.2. Under the hypotheses of Theorem 4.6.1, t_n may be chosen as the first of the numbers $\alpha^0, \alpha^1, \alpha^2, \ldots$ satisfying Equation 4.6.1, and then $\{x_n\}$ is a criticizing sequence,

$$f(x_n) - f(x_{n+1}) \geq \lambda_n d(\langle -\nabla f(x_n), p_n \rangle)$$

where

$$\lambda_n = 1 \quad \text{if} \quad t_n = 1, \lambda_n = \alpha s[(1 - \delta)\langle -\nabla f(x_n), p_n \rangle]$$

if $t_n \neq 1$.

Proof: As in the previous theorem, $t_n = 1$ yields no problem. In the other case, we have $x_{n+1} = x_n + \alpha^j p_n$, $j \geq 1$. Let $x_n' = x_n + \alpha^{j-1} p_n$. Then we have

$$f(x_n) - f(x_n') < \| x_n' - x_n \| \, d(\langle -\nabla f(x_n), p_n \rangle)$$
$$f(x_n) - f(x_{n+1}) \geq \| x_{n+1} - x_n \| \, d(\langle -\nabla f(x_n), p_n \rangle)$$

Therefore,

$$f(x_{n+1}) - f(x_n') < (1 - \alpha) \| x_n' - x_n \| \, d(\langle -\nabla f(x_n), p_n \rangle)$$

We can write

$$f(x_{n+1}) - f(x_n') = \langle \nabla f[\lambda_n x_n' + (1 - \lambda_n) x_{n+1}], x_{n+1} - x_n' \rangle$$

This leads to

$$\langle \nabla f[\lambda_n x_n' + (1 - \lambda_n) x_{n+1}], p_n \rangle > -d(\langle -\nabla f(x_n), p_n \rangle) \geq -\delta \langle -\nabla f(x_n), p_n \rangle$$

Hence

$$\begin{aligned} &\| \nabla f[\lambda_n x_n' + (1 - \lambda_n) x_{n+1}] - \nabla f(x_n) \| \\ &\quad \geq \langle \nabla f[\lambda_n x_n' + (1 - \lambda_n) x_{n+1}] - \nabla f(x_n), p_n \rangle \\ &\quad > (1 - \delta) \langle -\nabla f(x_n), p_n \rangle \end{aligned} \tag{4.6.5}$$

We then have

$$\| x_{n+1} - x_n \| \geq \alpha \| \lambda_n x_n' + (1 - \lambda_n) x_{n+1} - x_n \| \geq \alpha s[(1 - \delta)\langle -\nabla f(x_n), p_n \rangle]$$

Therefore, from this and from Equation 4.6.1 we have

$$f(x_n) - f(x_{n+1}) \geq \alpha s[(1 - \delta)\langle -\nabla f(x_n), p_n\rangle]d(\langle -\nabla f(x_n), p_n\rangle)$$

which implies that $\langle -\nabla f(x_n), p_n\rangle \longrightarrow 0$. Q.E.D.

In particular, one can consider this algorithm with $d(t) = \delta t$ and, instead of Equation 4.6.2, the stronger condition

$$g(x_n, t_n, p_n) \leq 1 - \delta$$

This method has been considered often [Goldstein (1964b, 1965, 1966, 1967)].

The two theorems above can be extended somewhat. For example, rather than demanding, in Theorem 4.6.1, that $||p_n|| = 1$, suppose we assume that

$$||p_n|| \geq d_1\left(\left\langle -\nabla f(x_n), \frac{p_n}{||p_n||}\right\rangle\right)$$

for some forcing function d_1, and that

$$\left\langle -\nabla f(x_n), \frac{p_n}{||p_n||}\right\rangle$$

tends to zero whenever

$$\frac{d(\langle -\nabla f(x_n), p_n\rangle)}{||p_n||}$$

tends to zero.

EXERCISE. Show that the latter condition immediately above is valid, for example, if $||p_n||$ is bounded above or $d(t) = qt, q \neq 0$.

Looking at the proof of Theorem 4.6.1, we see that under these conditions Equation 4.6.3 with $t_n = 1$ becomes

$$f(x_n) - f(x_{n+1}) \geq d(\langle -\nabla f(x_n), p_n\rangle) = d\left(\left\langle -\nabla f(x_n), \frac{p_n}{||p_n||}\right\rangle ||p_n||\right)$$

so that either

$$\left\langle -\nabla f(x_n), \frac{p_n}{||p_n||}\right\rangle \quad \text{or} \quad ||p_n|| \geq d_1\left(\left\langle -\nabla f(x_n), \frac{p_n}{||p_n||}\right\rangle\right)$$

must tend to zero, yielding $||\nabla f(x_n)|| \longrightarrow 0$. For $t_n \in (0, 1)$, Equation 4.6.4 becomes instead

$$t_n ||p_n|| \geq s\left[\frac{d(\langle -\nabla f(x_n), p_n\rangle)}{||p_n||}\right]$$

and thus

$$f(x_n) - f(x_{n+1}) \geq s\left[\frac{d(\langle -\nabla f(x_n), p_n\rangle)}{\|p_n\|}\right]\frac{d(\langle -\nabla f(x_n), p_n\rangle)}{\|p_n\|}$$

which implies that

$$\frac{d(\langle -\nabla f(x_n), p_n\rangle)}{\|p_n\|}$$

and thereby $\|\nabla f(x_n)\|$ tends to zero. Thus we have proved the following corollary.

COROLLARY 4.6.1. Theorem 4.6.1 is valid [except for the bound on $f(x_n) - f(x_{n+1})$] with the assumption that $\|p_n\| = 1$ being replaced by

1. $\|p_n\| \geq d_1\left(\left\langle -\nabla f(x_n), \frac{p_n}{\|p_n\|}\right\rangle\right)$;

and

2. $\left\langle -\nabla f(x_n), \frac{p_n}{\|p_n\|}\right\rangle \longrightarrow 0$ whenever $\frac{d(\langle -\nabla f(x_n), p_n\rangle)}{\|p_n\|} \longrightarrow 0$

Looking next at Theorem 4.6.2, we see that the case $t_n = 1$ follows as above. For $t_n \in (0, 1)$, Equation 4.6.5 becomes

$$\|p_n\| \|\nabla f[\lambda_n x_n' + (1 - \lambda_n)x_{n+1}] - \nabla f(x_n)\| \geq (1 - \delta)\langle -\nabla f(x_n), p_n\rangle$$

and thence

$$\|x_{n+1} - x_n\| \geq \alpha s\left[(1 - \delta)\left\langle -\nabla f(x_n), \frac{p_n}{\|p_n\|}\right\rangle\right]$$

and

$$f(x_n) - f(x_{n+1}) \geq \alpha s\left[(1 - \delta)\left\langle -\nabla f(x_n), \frac{p_n}{\|p_n\|}\right\rangle\right]\frac{d(\langle -\nabla f(x_n), p_n\rangle)}{\|p_n\|}$$

Thus we have proved the following two corollaries.

COROLLARY 4.6.2. Theorem 4.6.2 is valid [except for the bound on $f(x_n) - f(x_{n+1})$] with the assumption that $\|p_n\| = 1$ being replaced by conditions 1 and 2 in Corollary 4.6.1.

COROLLARY 4.6.3 [Armijo (1966)]. The conclusions of Theorems 4.6.1 and 4.6.2 are valid for the method defined by $p_n \equiv -\nabla f(x_n)$ and $d(t) = \delta t$, $\delta \in (0, \frac{1}{2}]$—that is, with t_n determined so that

$$f(x_n) - f(x_n - t_n \nabla f(x_n)) \geq \delta \|\nabla f(x_n)\|^2$$

Also, $\|x_{n+1} - x_n\| \longrightarrow 0$.

Proof: We may take $d_1(t) \equiv t$ for condition 1 of Corollary 4.6.1. For condition 2,

$$\frac{d(\langle -\nabla f(x_n), p_n \rangle)}{\| p_n \|} = \delta \| \nabla f(x_n) \|$$

$$\| x_{n+1} - x_n \| = t_n \| \nabla f(x_n) \| \leq \| \nabla f(x_n) \| \longrightarrow 0$$

Q.E.D.

In all of the above, note that if $\| \nabla f(x_n) \| \geq \epsilon \| p_n \|$, $\epsilon > 0$, then $\| x_{n+1} - x_n \| \longrightarrow 0$, since t_n is bounded; this is true in particular for $p_n = -\nabla f(x_n)$, as we saw above.

General references: Altman (1966a), Elkin (1968), Goldstein (1964b, 1965, 1966, 1967).

4.7. SEARCH METHODS ALONG THE LINE

In actual computation it is of course necessary to deal with discrete data; this means, for example, that one cannot generally minimize $f(x_n + tp_n)$ over all $t \geq 0$ but only over some discrete set of t-values. In this section we shall indicate how, in some cases, we can guarantee convergence for practical, computationally convenient choices of step size.

For theoretical analysis, we shall restrict ourselves to *strictly unimodal* functions—that is, to those that have a unique minimizing point along each straight line; from Section 1.5 we know that this is equivalent to strong quasi-convexity.

EXERCISE. Prove the equivalence of strict unimodality and strong quasi-convexity as asserted above.

This equivalence implies that if we have three t-values $t_1 < t_2 < t_3$ such that $f(x + t_2 p) < f(x + t_1 p)$ and $f(x + t_2 p) < f(x + t_3 p)$, then $f(x + tp)$ is minimized at a value of t between t_1 and t_3.

EXERCISE. Prove the preceding assertion concerning the location of the t-value minimizing the strictly unimodal function $f(x + tp)$.

We combine this fact with Theorem 4.4.2 for $\alpha_n \equiv \alpha = 0$ to prove the following.

THEOREM 4.7.1. Let f be strongly quasi-convex and bounded below on $W(x_0)$, let ∇f be uniformly continuous on $W(x_0)$, and let $p_n = p_n(x_n)$ define an admissible direction sequence. Suppose that for each n there are values $t_{n,1}, t_{n,2}, \ldots, t_{n,k_n}, t_{n,k_n+1}$ such that

$$f(x_n) > f(x_n + t_{n,1}p_n) > \cdots > f(x_n + t_{n,k_n}p_n) \leq f(x_n + t_{n,k_n+1}p_n)$$

and such that

$$\frac{t_{n,k_n-1}}{t_{n,k_n+1}} \geq \lambda > 0$$

for some constant λ. Then either $t_n \equiv t_{n,k_n-1}$ or $t_n \equiv t_{n,k_n}$ and $x_{n+1} = x_n + t_n p_n$ makes $\{x_n\}$ a criticizing sequence with

$$f(x_n) - f(x_{n+1}) \leq \lambda s(c\gamma_n)(1 - c)\gamma_n$$

for all c in (0, 1), with

$$\gamma_n \equiv \left\langle -\nabla f(x_n), \frac{p_n}{\|p_n\|} \right\rangle$$

Proof: The point t'_n providing the first local minimum for $f(x_n + tp_n)$ must satisfy $t_{n,k_n-1} < t_n < t_{n,k_n+1}$. Therefore, $t_{n,k_n-1} = \lambda_n t'_n$ where

$$\lambda_n \equiv \frac{t_{n,k_n-1}}{t'_n} \geq \frac{t_{n,k_n-1}}{t_{n,k_n+1}} \geq \lambda$$

and $\lambda_n \leq 1$. Thus Theorem 4.4.2 with $d(t) \equiv \lambda$ implies our theorem for $t_n \equiv t_{n,k_n-1}$. Since

$$f(x_n + t_{n,k_n}p_n) < f(x_n + t_{n,k_n-1}p_n)$$

part B of Theorem 4.2.4 proves the result for $t_n = t_{n,k_n}$ with $\beta = 1$. Q.E.D.

COROLLARY 4.7.1. Under the hypotheses of Theorem 4.7.1, if in addition $t_{n,i+1} - t_{n,i} = h_n$ for all i, then $k_n \geq 2$ is sufficient to guarantee that $t_n \equiv (k_n - 1)h_n$ or $t_n \equiv k_n h_n$ will make $\{x_n\}$ a criticizing sequence.

Proof: In this case,

$$\frac{t_{n,k_n-1}}{t_{n,k_n+1}} = \frac{k_n - 1}{k_n + 1} \geq \frac{1}{3} \equiv \lambda$$

Q.E.D.

COROLLARY 4.7.2 [Céa (1969)]. Under the hypotheses of Theorem 4.7.1, if in addition

$$f(x_n + 2h'_n p_n) > f(x_n + h'_n p_n) \leq f(x_n + \tfrac{1}{2}h'_n p_n) < f(x_n)$$

for some $h'_n > 0$, then $t_n \equiv h'_n$ makes $\{x_n\}$ a criticizing sequence.

Proof: If

$$f(x_n + \tfrac{3}{2}h'_n p_n) \leq f(x_n + h'_n p_n)$$

we have an example of Corollary 4.7.1 with $k_n = 3$ and $h_n = h'_n/2$. On the other hand, if

$$f(x_n + \tfrac{3}{2}h'_n p_n) > f(x_n + h'_n p_n)$$

we have an example of Corollary 4.7.1 with $k_n = 2$ and $h_n = h'_n/2$. Q.E.D.

We shall combine these results into a single algorithm in a moment; since a simplification is possible if f is actually convex, we derive one more result first.

THEOREM 4.7.2 [Céa (1969)]. In addition to the hypotheses of Theorem 4.7.1, suppose that f is convex and that for all n we have $h_n > 0$ such that

$$f(x_n + h_n p_n) \leq f(x_n + 2h_n p_n) \leq f(x_n)$$

Then $t_n \equiv h_n$ makes $\{x_n\}$ a criticizing sequence and

$$f(x_n) - f(x_{n+1}) \geq \tfrac{1}{2}s(c\gamma_n)(1 - c)\gamma_n$$

for all c in (0, 1), with

$$\gamma_n \equiv \left\langle -\nabla f(x_n), \frac{p_n}{\|p_n\|}\right\rangle$$

Proof: The point t'_n providing the global minimum for $f(x_n + tp_n)$ must satisfy $0 \leq t'_n \leq 2h_n$ and of course $t_n \equiv t'_n$ would yield a criticizing sequence. Since f is convex, for $0 \leq t \leq h_n$ we have

$$\begin{aligned} f(x_n + tp_n) &\geq 2f(x_n + h_n p_n) - f(x_n + 2h_n p_n) \\ &\quad + \frac{f(x_n + 2h_n p_n) - f(x_n + h_n p_n)}{h_n} t \\ &\geq 2f(x_n + h_n p_n) - f(x_n + 2h_n p_n) \\ &\geq 2f(x_n + h_n p_n) - f(x_n) \end{aligned}$$

while arguing similarly for $h_n \leq t \leq 2h_n$ we deduce

$$f(x_n + tp_n) \geq 2f(x_n + h_n p_n) - f(x_n)$$

Setting $t'_n = t$ thus gives

$$f(x_n + t'_n p_n) \geq 2f(x_n + h_n p_n) - f(x_n)$$

and therefore

$$f(x_n) - f(x_n + h_n p_n) \geq \tfrac{1}{2}[f(x_n) - f(x_n + t'_n p_n)]$$

The theorem follows from part B of Theorem 4.2.4 with $\beta = \frac{1}{2}$. Q.E.D.

We can now give a practical algorithm of a search type to locate a suitable value of t_n. We assume that the algorithm is entered with a point x_n, direction p_n, and a number $h > 0$ given. We write in a pseudo-ALGOL language for convenience.

Search routine [Céa (1969)]

```
            if f(x_n + hp_n) < f(x_n) then go to first;
reduce:     h ← h/2;
            if f(x_n + hp_n) ≥ f(x_n) then go to reduce;
            if f(x_n + (h/2)p_n) ≥ f(x_n + hp_n) then EXIT FROM ROU-
                TINE NOW WITH t_n = h;
            if f IS CONVEX then EXIT FROM ROUTINE NOW
                WITH t_n = h/2;
            h ← h/2;
loop:       while f(x_n + (h/2)p_n) < f(x_n + hp_n) do h ← h/2;
            EXIT FROM ROUTINE NOW WITH t_n = h;
first:      if f(x_n + 2hp_n) ≥ f(x_n + hp_n) then go to oldway;
            t ← 2h;
change:     while f(x_n + (t + h)p_n) < f(x_n + tp_n) do t ← t + h;
            EXIT FROM ROUTINE NOW WITH t_n = t;
oldway:     if f IS CONVEX and f(x_n + 2hp_n) ≤ f(x_n) then EXIT
                FROM ROUTINE NOW WITH t_n = h;
            go to loop;
```

It would also of course be possible to move more rapidly by replacing the one line with the label "change" with the line

```
change:     while f(x_n + 2tp_n) < f(x_n + tp_n) do t ← 2t;
```

THEOREM 4.7.3. Let f be strongly quasi-convex and bounded below on $W(x_0)$, let ∇f be uniformly continuous on $W(x_0)$, and let $p_n = p_n(x_n)$ define an admissible direction sequence. Let t_n be determined by the above search routine. Then $\{x_n\}$ is a criticizing sequence.

EXERCISE. Supply the *Proof* to Theorem 4.7.3.

4.8. SPECIALIZATION TO STEEPEST DESCENT

The general gradient-type methods we have been discussing are generalizations of the original method of steepest descent [Cauchy (1847)]. In that special case we suppose that E is a Hilbert space and we let $p_n(x_n) = -\nabla f(x_n)$, which clearly is an admissible sequence of directions. Thus all of the theorems we have developed in this chapter yield corollaries when applied to this method. In some cases, however, one can go further for the original steepest-descent method and give estimates on the rate of convergence. The next chapter, for example, will contain, as a by-product, convergence estimates for the steepest-descent directions selecting t_n as in Section 4.3 and Section 4.4. Therefore, at this point we shall only demonstrate the results obtainable for selecting t_n from a simple interval along the line.

THEOREM 4.8.1. Suppose f is a twice-differentiable functional on a Hilbert space E and that $mI \leq f''_x \leq MI$ for $0 < m \leq M < \infty$ for all x. Let $\delta_1 > 0$ and $\delta_2 > 0$ be chosen and choose t_n to satisfy

$$\delta_1 \leq t_n \leq \frac{2}{M} - \delta_2$$

and set $x_{n+1} = x_n - t_n \nabla f(x_n)$. Then $x_n \longrightarrow x^*$, the unique point minimizing f over E, starting at any x_0. Given any $\epsilon > 0$ there exists an N such that for $n \geq N$,

$$\| x^* - x_{n+1} \| \leq \| x^* - x_n \| (\lambda_n + \epsilon), \qquad \lambda_n = \max(|1 - t_n m|, |1 - t_n M|)$$

The error estimate is best when

$$t_n \equiv t^* = \frac{M + m}{2} \quad \text{for all} \quad n$$

In this case, then, x_n converges faster than any geometric series with ratio greater than $(M - m)/(M + m)$.

Proof: By Theorem 1.4.4, $\lim_{\|x\| \to \infty} f(x) = \infty$, so we may restrict the problem to a bounded set—that is, $W(x_0)$ is bounded for each x_0. Since

$$f(x) \geq f(0) - \| \nabla f(0) \| \| x \| + \frac{m}{2} \| x \|^2$$

f is bounded below. Since $\| f''_x \| \leq M$, ∇f is Lipschitz-continuous with Lipschitz constant M; Theorem 4.5.1 with $A_1 = A_2 = 1$ then says that $\{x_n\}$ is a criticizing sequence. From Theorem 4.2.1 it follows that $\{x_n\}$ is a minimizing sequence and then from Theorem 1.6.3 we conclude that $x_n \longrightarrow x^*$, the unique point in $W(x_0)$ and E minimizing f. Now we wish to consider the convergence rate:

$$\begin{aligned} x^* - x_{n+1} &= x^* - x_n + t_n\nabla f(x_n) = x^* - x_n + t_n f''_{x^*}(x_n - x^*) \\ &\quad + t_n[\nabla f(x_n) - f''_{x^*}(x_n - x^*)] \\ &= [I - t_n f''_{x^*}](x^* - x_n) + t_n[\nabla f(x_n) - \nabla f(x^*) - f''_{x^*}(x_n - x^*)] \end{aligned}$$

since $\nabla f(x^*) = 0$. By the definition of f''_{x^*},

$$\nabla f(x_n) - \nabla f(x^*) - f''_{x^*}(x_n - x^*) = \|x_n - x^*\| \, w(\|x_n - x^*\|)$$

where $\lim_{s \to 0} w(s) = 0$. Thus

$$\|x^* - x_{n+1}\| \leq \|x^* - x_n\| \left[\|I - t_n f''_{x^*}\| + \frac{2}{M} w(\|x_n - x^*\|)\right]$$

Therefore,

$$\begin{aligned} \|x^* - x_{n+1}\| &\leq \|x^* - x_n\| \\ &\quad \times \left[\max(|1 - t_n m|, |1 - t_n M|) + \frac{2}{M} w(\|x_n - x^*\|)\right] \end{aligned}$$

Given $\epsilon < 0$, then for large $n \geq N$,

$$\frac{2}{M} w(\|x_n - x^*\|) \leq \epsilon$$

which gives

$$\|x^* - x_{n+1}\| \leq \|x^* - x_n\| (\lambda_n + \epsilon)$$

with $\lambda_n = \max(|1 - t_n m|, |1 - t_n M|) < 1$. Each λ_n can be minimized by choosing

$$t_n = t^* = \frac{m + M}{2} = \delta_1 = \delta_2$$

in which case

$$\lambda_n = \frac{M - m}{M + m}$$

Q.E.D.

Computer programs implementing steepest descent in $\mathbb{R}^l$ may be found in Whitley (1962), Wasscher (1963).

General references: Kantorovich (1948), Levitin–Poljak (1966a).

4.9. STEP-SIZE ALGORITHMS FOR CONSTRAINED PROBLEMS

Although we shall not attempt to examine the many various kinds of iterative methods for treating problems with constraints, we do wish to see to what extent the methods of the previous sections can be modified for use on these problems. We remarked in Section 1.4 that a necessary condition for a point x^* to provide a minimum for a differentiable function f over a convex set C is that $\langle x - x^*, -\nabla f(x^*)\rangle \leq 0$ for all $x \in C$—that is, that all directions leading into C make obtuse angles with the direction of steepest descent, implying that f is nondecreasing in every direction pointing into C. If this condition is not satisfied at a point x_0, the natural step would be to find a direction $p_0 = x'_0 - x_0$, $x'_0 \in C$, with $\langle p_0, -\nabla f(x_0)\rangle > 0$ and move to a new point $x_1 \in C$ along this direction. One might hope to find such a point x'_0 by "projecting" the gradient direction or some other admissible direction onto the set C. Unfortunately, an arbitrary direction making a strict *acute* angle with $-\nabla f(x_0)$ can "project" into a direction p_0 making an *obtuse* angle with $-\nabla f(x_0)$, leading to no decrease in the value for f. Thus when we consider "projection" methods, we shall have to deal directly with $-\nabla f(x_0)$. First, however, let us consider in general what happens if one uses feasible directions p_n [Topkis–Veinott (1967), Zangwill (1969) Zoutendijk (1960)].

DEFINITION 4.9.1. A sequence of directions $p_n = p_n(x_n)$ is called *feasible* if and only if $p_n = x'_n - x_n$ where $\lambda x_n + (1 - \lambda)x'_n \in C$ for all $\lambda \in [0, 1]$, $x'_n \neq x_n$, and $\langle p_n, -\nabla f(x_n)\rangle \geq 0$.

For unconstrained problems, with $\|p_n\| = 1$, if $\langle p_n, \nabla f(x_n)\rangle \longrightarrow 0$ implied $\|\nabla f(x_n)\| \longrightarrow 0$, then the direction sequence was a "good" one, in the sense that reasonable step-size algorithms yielded criticizing sequences. For the constrained problem we shall similarly deduce $\langle p_n, \nabla f(x_n)\rangle \longrightarrow 0$ for many methods, so the problem will be to choose directions to avoid "jamming" [Zangwill (1969)] or "zigzagging" [Zoutendijk (1960)] so that "in the limit" the condition "$\langle \nabla f(x), p(x)\rangle = 0$" is a necessary or sufficient optimality criterion; in Section 4.10 we give some examples of directions for which $\langle \nabla f(x_n), p_n\rangle \longrightarrow 0$ is a useful condition.

Since we have already analyzed step-size algorithms for unconstrained problems, we would like to make use of those results here. Unfortunately, a step size generated by some method ignoring constraints [e.g., finding the first local minimum for $f(x_n + tp_n)$] may yield an x_{n+1} outside the constraint set C; in this case one might consider proceeding as far along p_n toward that point as possible while staying within C. As we shall see, in many cases this is a good strategy. First we recall that for the algorithms we have considered for unconstrained problems, the t_n that was determined always guaranteed that

$$f(x_n) - f(x_n + t_n p_n) \geq d\left(\left\langle -\nabla f(x_n), \frac{p_n}{\|p_n\|}\right\rangle\right)$$

for some forcing function $d(t)$, depending on the method; we can analyze all such algorithms of this type. As usual, of course, we remark that any different x'_{n+1} may be used satisfying $f(x_n) - f(x'_{n+1}) \geq \beta[f(x_n) - f(x_{n+1})]$ for fixed $\beta > 0$ as in Theorem 4.2.4; we shall not continually repeat this obvious fact.

THEOREM 4.9.1. Let the convex functional f be bounded below on the bounded convex set C and, for some x_0 in C, let the set $\{x; f(x) \leq f(x_0)\}$ be bounded; let $p_n = p_n(x_n)$ define a feasible direction sequence and let $\|\nabla f(x)\|$ be uniformly bounded for $x \in C \cap \{x; f(x) \leq f(x_0)\}$. Let the numbers t_n^u be some steps satisfying

$$f(x_n) - f(x_n + t_n^u p_n) \geq d\left(\left\langle -\nabla f(x_n), \frac{p_n}{\|p_n\|}\right\rangle\right)$$

Let $x_{n+1} = x_n + t_n p_n$ where $t_n = t_n^u$ if $x_n + t_n^u p_n \in C$ and $t_n = t_n^l \geq \epsilon > 0$ with $s_n + t_n^l p_n \in C$ otherwise. Then $\langle \nabla f(x_n), p_n \rangle \to 0$.

Proof: If $t_n = t_n^u$, then

$$f(x_n) - f(x_{n+1}) \geq d\left(\left\langle -\nabla f(x_n), \frac{p_n}{\|p_n\|}\right\rangle\right)$$

If

$$t_n = t_n^l \quad \text{and} \quad f(x_n + t_n^l p_n) \leq f(x_n + t_n^u p_n)$$

then also

$$f(x_n) - f(x_{n+1}) \geq d\left(\left\langle -\nabla f(x_n), \frac{p_n}{\|p_n\|}\right\rangle\right)$$

We consider the final case of

$$t_n = t_n^l \quad \text{and} \quad f(x_n + t_n^l p_n) > f(x_n + t_n^u p_n)$$

Since f is convex and $t_n^u > t_n^l$, we have

$$f(x_n + t_n^l p_n) = f\left(x_n + \frac{t_n^l}{t_n^u} t_n^u p_n\right) \leq \left(1 - \frac{t_n^l}{t_n^u}\right) f(x_n) + \frac{t_n^l}{t_n^u} f(x_n + t_n^u p_n)$$

and thus

$$f(x_n) - f(x_n + t_n^l p_n) \geq \frac{t_n^l \|p_n\|}{t_n^u \|p_n\|} [f(x_n) - f(x_n + t_n^u p_n)]$$

Since $\{x; f(x) \leq f(x_0)\}$ is bounded, there is a K such that $\| t_n^u p_n \| \leq K$ and, therefore,

$$f(x_n) - f(x_{n+1}) \geq t_n^l \| p_n \| \frac{1}{K} d\left(\left\langle -\nabla f(x_n), \frac{p_n}{\| p_n \|}\right\rangle\right)$$

Since $t_n^l \geq \epsilon > 0$ and $\| \nabla f(x_n) \|$ is uniformly bounded and

$$\langle -\nabla f(x_n), p_n \rangle = \left\langle -\nabla f(x_n), \frac{p_n}{\| p_n \|}\right\rangle \| x_n' - x_n \| \leq K\left\langle -\nabla f(x_n), \frac{p_n}{\| p_n \|}\right\rangle$$

then from the three inequalities for $f(x_n) - f(x_{n+1})$ we deduce that $\langle \nabla f(x_n), p_n \rangle \longrightarrow 0$. Q.E.D.

Remarks. Since $p_n = x_n' - x_n$ for some $x_n' \in C$, $t_n' = 1$ is always allowed, and so certainly $t_n^l \geq \epsilon > 0$ is possible; in particular, $t_n^l = \max \{t; x_n + tp_n \in C\}$ is possible. If C is itself bounded, by modifying and redefining f outside of C we can generally guarantee that $\{x; f(x) \leq f(x_0)\}$ is bounded.

For the step-size algorithms studied for the unconstrained problems, it is possible to eliminate the hypotheses that f is convex in the above theorem. In general, the proofs of these facts follow the arguments for the unconstrained case, so we shall be rather brief; first we state a theorem similar to Theorem 4.2.4, again using the reverse modulus of continuity $s(t)$ defined in Definition 4.2.3.

THEOREM 4.9.2. Let f be bounded below on $W(x_0)$, let ∇f be uniformly continuous and uniformly bounded on $W(x_0)$, let C be convex and bounded, and let $p_n = p_n(x_n)$ define a feasible direction sequence for C. Let there exist functions $c_1(t)$ and $c_2(t)$ such that $c_1(t)$ and $t - c_2(t)$ are forcing functions. Let t_n^u be step sizes such that

$$c_1\left(\left\langle -\nabla f(x_n), \frac{p_n}{\| p_n \|}\right\rangle\right) \leq t_n^u \| p_n \| \leq s\left[c_2\left(\left\langle -\nabla f(x_n), \frac{p_n}{\| p_n \|}\right\rangle\right)\right]$$

and let t_n^l be step sizes such that

$$x_n + t_n^l p_n \in C \quad \text{and} \quad t_n^l \| p_n \| \geq d_1(\| p_n \|) d_2\left(\left\langle -\nabla f(x_n), \frac{p_n}{\| p_n \|}\right\rangle\right)$$

for two forcing functions $d_1(t)$ and $d_2(t)$.

1. If we set $t_n = t_n^u$ if $x_n + t_n^u p_n \in C$ and $t_n = t_n^l$ otherwise, with $x_{n+1} = x_n + t_n p_n$, we conclude that $\langle \nabla f(x_n), p_n \rangle \longrightarrow 0$.
2. If t_n' is chosen as is t_n above in 1 from $t_n^{u'}$ and $t_n^{l'}$ and x_{n+1} is chosen so that

$$f(x_n) - f(x_{n+1}) \geq \beta[f(x_n) - f(x_n + t_n' p_n)]$$

for a fixed $\beta > 0$, then $\langle \nabla f(x_n), p_n \rangle \longrightarrow 0$.

Proof: As in Theorem 4.2.4, we easily find

$$f(x_n) - f(x_n + t_n^u p_n) \geq t_n^u \|p_n\| [\gamma_n - c_2(\gamma_n)] \geq c_1(\gamma_n)[\gamma_n - c_2(\gamma_n)]$$

where

$$\gamma_n = \left\langle -\nabla f(x_n), \frac{p_n}{\|p_n\|} \right\rangle$$

If $t_n = t_n^u$, we then have

$$f(x_n) - f(x_{n+1}) \geq c_1(\gamma_n)[\gamma_n - c_2(\gamma_n)]$$

If $t_n = t_n^l$, then $t_n^l \|p_n\| < t_n^u \|p_n\|$, and arguing as for t_n^u we get

$$\begin{aligned} f(x_n) - f(x_n + t_n^l p_n) &\geq t_n^l \|p_n\| [\gamma_n - c_2(\gamma_n)] \\ &\geq d_2(\gamma_n) d_1(\|p_n\|)[\gamma_n - c_2(\gamma_n)] \end{aligned}$$

Thus $\gamma_n \longrightarrow 0$ or $\|p_n\| \longrightarrow 0$; since $\|p_n\| = \|x_n' - x_n\|$ and $\|\nabla f(x_n)\|$ are bounded, this gives $\langle \nabla f(x_n), p_n \rangle \longrightarrow 0$. Part 2 follows easily from the estimates of part 1. Q.E.D.

Remark. The convexity hypothesis on C can be removed easily here and in what follows. As in the unconstrained case, this general theorem makes it easy to analyze the convergence of many step-size algorithms.

THEOREM 4.9.3. Let f be bounded below on $W(x_0)$, let ∇f be uniformly continuous and uniformly bounded on $W(x_0)$, let C be convex and bounded, and let $p_n = p_n(x_n)$ define a feasible direction sequence for C. For numbers $\alpha_n \in [0, \alpha]$ with $\alpha < 1$, choose t_n such that $x_{n+1} \equiv x_n + t_n p_n \in C$ and

$$f(x_n + t_n p_n) - \alpha_n t_n \langle \nabla f(x_n), p_n \rangle \leq f(x_n + t p_n) - \alpha_n t \langle \nabla f(x_n), p_n \rangle$$

for all $t \geq 0$ such that $x_n + t p_n \in C$. Then $\langle \nabla f(x_n), p_n \rangle \longrightarrow 0$.

Proof: By part 1 of Theorem 4.9.2 with $c_1(t) \equiv s(ct)$ and $c_2(t) \equiv ct$ for fixed $c \in (0, 1 - \alpha)$, $d_1(t) \equiv t$, and $d_2(t) \equiv 1$, the algorithm with t_n' determined from

$$t_n^{u'} \|p_n\| \equiv s\left[c\left\langle -\nabla f(x_n), \frac{p_n}{\|p_n\|} \right\rangle\right] \quad \text{and} \quad t_n^{l'} \equiv 1$$

gives the desired convergence. We know that either

$$\langle \nabla f(x_n + t_n p_n), p_n \rangle - \alpha_n \langle \nabla f(x_n), p_n \rangle = 0$$

or

$$t_n = \sup \{t; x_n + t p_n \in C\}$$

In the first case, arguing as in the proof of Theorem 4.3.1, we find $t_n \geq t_n^{u'}$, which implies $t'_n = t_n^{u'}$ and thus $t_n \geq t'_n$; in the second case, clearly $t_n \geq t'_n$. In either case, by the defining property of t_n and the fact that $t_n \geq t'_n$, we have

$$\begin{aligned} f(x_n) - f(x_n + t_n p_n) &\geq f(x_n) - f(x_n + t'_n p_n) + \alpha_n(t_n - t'_n)\langle -\nabla f(x_n), p_n \rangle \\ &\geq f(x_n) - f(x_n + t'_n p_n) \end{aligned}$$

so that the theorem follows from part 2 of Theorem 4.9.2 with $\beta \equiv 1$. Q.E.D.

EXERCISE. Fill in the details in the above *Proof.*

Remark. Setting $\alpha_n \equiv \alpha = 0$ yields the usual method.

THEOREM 4.9.4. Let f and C be as in Theorem 4.9.3 and let C be norm-closed. Let t_n be either: (1) the smallest positive t_n providing a local minimum for

$$f(x_n + t_n p_n) - \alpha_n t \langle \nabla f(x_n), p_n \rangle$$

over the set of t such that $x_n + tp_n \in C$, $t \geq 0$; or (2) the first positive root r of

$$\langle \nabla f(x_n + tp_n), p_n \rangle - \alpha_n \langle \nabla f(x_n), p_n \rangle = 0$$

if $x_n + rp_n \in C$, otherwise

$$t_n = \sup \{t; x_n + tp_n \in C\}$$

or (3) the following:

$$\begin{aligned} t_n = \sup \{t; &\langle \nabla f(x_n + \tau p_n), p_n \rangle - \alpha_n \langle \nabla f(x_n), p_n \rangle \leq 0 \\ &\text{for} \quad 0 \leq \tau \leq t, \quad x_n + tp_n \in C\} \end{aligned}$$

We assume $0 \leq \alpha_n \leq \alpha < 1$. Let $x_{n+1} = x_n + t_n p_n$; then $\langle \nabla f(x_n), p_n \rangle \to 0$.

Proof: The theorem follows from Theorem 4.9.2 precisely as does Theorem 4.4.1 from Theorem 4.2.4. Q.E.D.

EXERCISE. Give the complete proof for the constrained-minimization analogue of Corollary 4.4.1—that is, for $\alpha_n \equiv \alpha \equiv 0$, in Theorem 4.9.4.

THEOREM 4.9.5. Let f and C be as in Theorem 4.9.4 and let $\hat{t}_n$ be determined as is t_n in that theorem. Let $t_n \equiv \lambda_n \hat{t}_n$ where

$$d\left(\left\langle -\nabla f(x_n), \frac{p_n}{\|p_n\|} \right\rangle\right) \leq \lambda_n \leq 1$$

for some forcing function $d(t)$, and let $x_{n+1} = x_n + t_n p_n$; then $\langle \nabla f(x_n), p_n \rangle \to 0$.

Proof: The theorem follows from Theorem 4.9.2 precisely as does Theorem 4.4.2 from Theorem 4.2.4 by using, for fixed $c \in (0, 1 - \alpha)$, $c_1(t) \equiv d(t)s(ct)$, $c_2(t) \equiv ct$,

$$t_n^{u'} \| p_n \| \equiv d\left(\left\langle -\nabla f(x_n), \frac{p_n}{\| p_n \|} \right\rangle\right) s\left(c\left\langle -\nabla f(x_n), \frac{p_n}{\| p_n \|} \right\rangle\right)$$

$$\begin{aligned} t_n^{l'} &= \lambda_n \sup \{t; x_n + tp_n \in C\} \\ &\geq \lambda_n \geq d\left(\left\langle -\nabla f(x_n), \frac{p_n}{\| p_n \|} \right\rangle\right) \end{aligned}$$

$d_1(t) \equiv t$, $d_2(t) \equiv d(t)$, $\beta \equiv 1$. Q.E.D.

EXERCISE. Fill in the details in the above *Proof.*

As in the unconstrained case, if ∇f is Lipschitz-continuous, while the above theorems define a range of t-values leading to $\langle \nabla f(x_n), p_n \rangle \longrightarrow 0$, it is possible to double the size of this range by a more careful analysis.

THEOREM 4.9.6. Let f be bounded below on C, let ∇f satisfy

$$\| \nabla f(x) - \nabla f(y) \| \leq L \| x - y \|$$

for x, y in C, and let $p_n = p_n(x_n)$ define a feasible direction sequence. Pick $\delta_1, \delta_2, \delta_3$ all greater than zero and let γ_n lie in

$$\left[\min\left(\delta_1, \frac{\delta_2 \| p_n \|^2}{-\langle \nabla f(x_n), p_n \rangle}\right), \frac{2}{L} - \delta_3\right]$$

for all n. For each n let $x_{n+1} = x_n + t_n p_n$ where t_n is defined via

$$t_n = \min\left(1, \frac{\gamma_n \langle -\nabla f(x_n), p_n \rangle}{\| p_n \|^2}\right)$$

Then $f(x_n)$ decreases to a limit. If $\| p_n \|$ is uniformly bounded—for example, if C is bounded—then

$$\lim_{n \to \infty} \langle \nabla f(x_n), p_n \rangle = 0$$

If $\| p_n \| \longrightarrow 0$ implies

$$\left\langle \nabla f(x_n), \frac{p_n}{\| p_n \|} \right\rangle \longrightarrow 0$$

then

$$\lim_{n \to \infty} \left\langle \nabla f(x_n), \frac{p_n}{\| p_n \|} \right\rangle = 0$$

Proof:

$$\begin{aligned} f(x_{n+1}) - f(x_n) &\le \langle \nabla f(x_n), x_{n+1} - x_n \rangle \\ &\quad + \int_0^1 \langle \nabla f(x_n + \lambda t_n p_n) - \nabla f(x_n), t_n p_n \rangle d\lambda \\ &\le \langle \nabla f(x_n), x_{n+1} - x_n \rangle + \frac{L}{2} t_n^2 \|p_n\|^2 \\ &\le -t_n \langle -\nabla f(x_n), p_n \rangle + \frac{L}{2} t_n^2 \|p_n\|^2 \end{aligned}$$

If

$$1 \le \gamma_n \frac{\langle -\nabla f(x_n), p_n \rangle}{\|p_n\|^2}$$

then $t_n = 1$, x_{n+1} is in C, and

$$\begin{aligned} f(x_{n+1}) - f(x_n) &\le \langle -\nabla f(x_n), p_n \rangle \left[-1 + \frac{L}{2} \frac{\|p_n\|^2}{\langle -\nabla f(x_n), p_n \rangle} \right] \\ &\le \langle -\nabla f(x_n), p_n \rangle \left[-1 + \frac{L\gamma_n}{2} \right] \le \frac{-\delta_3 L}{2} \langle -\nabla f(x_n), p_n \rangle \le 0 \end{aligned}$$

If, however,

$$1 > t_n = \gamma_n \frac{\langle -\nabla f(x_n), p_n \rangle}{\|p_n\|^2}$$

then x_{n+1} is in C and

$$\begin{aligned} f(x_{n+1}) - f(x_n) &\le \gamma_n \frac{\langle -\nabla f(x_n), p_n \rangle^2}{\|p_n\|^2} + \frac{L}{2} \|p_n\|^2 \gamma_n^2 \frac{\langle -\nabla f(x_n), p_n \rangle^2}{\|p_n\|^4} \\ &\le \frac{\langle -\nabla f(x_n), p_n \rangle^2}{\|p_n\|^2} \left[\frac{\gamma_n^2 L}{2} - \gamma_n \right] \\ &\le \text{ either } \frac{-\delta_1 \delta_3 L}{2} \frac{\langle -\nabla f(x_n), p_n \rangle^2}{\|p_n\|^2} \\ &\qquad \text{or } \frac{-\delta_2 \delta_3 L}{2} \langle -\nabla f(x_n), p_n \rangle \end{aligned}$$

In either case, $f(x_{n+1}) - f(x_n) \le 0$ and $f(x_n)$ decreases to a limit. If $\|p_n\| = \|x_n' - x_n\|$ is bounded, then from the three inequalities bounding the decrease in f we obtain a $\delta > 0$ such that

$$f(x_n) - f(x_{n+1}) \ge \delta \langle -\nabla f(x_n), p_n \rangle^r$$

for either $r = 1$ or $r = 2$, which implies $\lim_{n \to \infty} \langle -\nabla f(x_n), p_n \rangle = 0$. Since

$$\left\langle \nabla f(x_n), \frac{p_n}{\|p_n\|} \right\rangle = \frac{\langle \nabla f(x_n), p_n \rangle}{\|p_n\|}$$

the final conclusion also follows. Q.E.D.

Other simple range theorems in terms of Lipschitz constants can be given analogous to Theorems 4.5.2 and 4.5.3 in the unconstrained case; we leave these as exercises and proceed to the more complex methods of Section 4.6.

We determine admissible values of t_n in terms of the *range function*

$$g(x, t, p) \equiv \frac{f(x) - f(x + tp)}{-t\langle \nabla f(x), p\rangle}$$

Given a feasible direction sequence defined by $p_n = p_n(x_n)$, for the moment assuming $\|p_n\| = 1$, a real number $\delta \in (0, \frac{1}{2}]$, and a forcing function $d(t) \leq \delta t$, we move from x_n to x_{n+1} as follows. If, for $t_n = 1$ and $x'_n = x_n + p_n$, we find

$$g(x_n, t_n, p_n) \geq \frac{d(\langle -\nabla f(x_n), p_n\rangle)}{\langle -\nabla f(x_n), p_n\rangle} \tag{4.9.1}$$

we set $x_{n+1} = x'_n$; otherwise, find t_n in $(0, 1)$ satisfying Equation 4.9.1 and also

$$|g(x_n, t_n, p_n) - 1| \geq \frac{d(\langle -\nabla f(x_n), p_n\rangle)}{\langle -\nabla f(x_n), p_n\rangle} \tag{4.9.2}$$

and set $x_{n+1} = x_n + t_n p_n \in C$ since $x_n + p_n \in C$. We observe that the algorithm is well defined. Since $g(x_n, 0, p_n) = 1$ and

$$1 - \frac{d(t)}{t} \geq \frac{d(t)}{t}$$

for all t, if we have

$$g(x_n, 1, p_n) < \frac{d(z)}{z} \quad \text{where} \quad z = \langle -\nabla f(x_n), p_n\rangle$$

then by continuity of $g(x_n, t, p_n)$ in t and the fact that $x_n + tp_n$ is in C for t in $[0, 1]$ since p_n is a feasible direction, there exists t_n in $(0, 1)$ with

$$\frac{d(z)}{z} \leq g(x_n, t_n, p_n) \leq 1 - \frac{d(z)}{z}$$

which certainly satisfies Equations 4.9.1 and 4.9.2.

THEOREM 4.9.7. Let f be bounded below on C, ∇f be uniformly continuous on C, and $p_n \equiv p_n(x_n)$ be a feasible direction sequence with $\|p_n\| = 1$. Let d be a forcing function with $d(t) \leq \delta t$ for δ in $(0, \frac{1}{2}]$. Let the algorithm described above be applied. Then $\lim_{n\to\infty} \langle \nabla f(x_n), p_n\rangle = 0$.

Proof: The proof is exactly the same as that for Theorem 4.6.1. Q.E.D.

For problems in which C is not the whole space—that is, in which there are constraints—the restriction $\|p_n\| = 1$ is unrealistic; the following corollary shows that it is not needed so long as p_n cannot be "too small" compared to how "near" one is to a solution.

COROLLARY 4.9.1. Under the hypotheses of Theorem 4.9.7, above, with the assumption $\|p_n\| = 1$ replaced by

1. $\|p_n\| \geq d_1\left(\left\langle -\nabla f(x_n), \frac{p_n}{\|p_n\|}\right\rangle\right)$ for a forcing function d_1, and
2. $\left\langle -\nabla f(x_n), \frac{p_n}{\|p_n\|}\right\rangle \longrightarrow 0$ whenever $\frac{d(\langle -\nabla f(x_n), p_n\rangle)}{\|p_n\|} \longrightarrow 0$,

it follows that

$$\lim_{n\to\infty} \left\langle \nabla f(x_n), \frac{p_n}{\|p_n\|}\right\rangle = 0$$

Proof: The proof is exactly the same as that for Corollary 4.6.1. Q.E.D.

For the direction algorithms we shall consider in Section 4.10 for constrained problems, it will not be necessary for us to have

$$\left\langle \nabla f(x_n), \frac{p_n}{\|p_n\|}\right\rangle \longrightarrow 0$$

The condition $\langle \nabla f(x_n), p_n\rangle \longrightarrow 0$ will be enough. Considering the comments prior to Corollaries 4.6.1 and 4.6.2, it is immediately clear that $\langle \nabla f(x_n), p_n\rangle \longrightarrow 0$ even without assumption 1 of Corollary 4.9.1. Thus we have the following corollary.

COROLLARY 4.9.2. Under the hypotheses of Theorem 4.9.7, above, with the hypothesis $\|p_n\| = 1$ replaced by

$$\left\langle -\nabla f(x_n), \frac{p_n}{\|p_n\|}\right\rangle \longrightarrow 0$$

whenever

$$\frac{d(\langle -\nabla f(x_n), p_n\rangle)}{\|p_n\|} \longrightarrow 0$$

it follows that $\langle \nabla f(x_n), p_n\rangle \longrightarrow 0$.

EXERCISE. Prove Corollary 4.9.2 in detail.

The algorithm above is not computational in that it may well be very difficult to locate a $t_n \in (0, 1)$ satisfying Equations 4.9.1 and 4.9.2; the algorithm of Theorem 4.6.2 and Corollary 4.6.2 works for the unconstrained problem as well, yielding the desired computational procedure. The proofs of the following results are exactly the same as those for Theorem 4.6.2 and Corollary 4.6.2.

THEOREM 4.9.8. Under the hypotheses of Theorem 4.9.7 or Corollary 4.9.1 or Corollary 4.9.2, t_n may be chosen as the first of the numbers α^0, $\alpha^1, \alpha^2, \ldots$ satisfying Equation 4.9.1 for a fixed $\alpha \in (0, 1)$; if $x_{n+1} = x_n + t_n p_n$, then $\langle \nabla f(x_n), p_n \rangle \longrightarrow 0$.

It is also quite clear that the search routine described in Section 4.7 works equally well on constrained problems so long as $x_n + hp_n \in C$ for the initial h and the increasing of t in the step labeled "change" is not allowed to force $x_n + tp_n$ outside of C.

EXERCISE. By proving the analogues to Theorems 4.7.1 and 4.7.2 and to Corollaries 4.7.1 and 4.7.2, prove Theorem 4.9.9 below.

THEOREM 4.9.9. Let f be strongly quasi-convex and bounded below on $W(x_0)$, let ∇f be uniformly continuous and uniformly bounded on $W(x_0)$, let C be convex and bounded, and let $p_n = p_n(x_n)$ define a feasible direction sequence for C. Let the search routine described below be used to determine t_n, where $x_n + hp_n \in C$, and set $x_{n+1} = x_n + t_n p_n$; then $\langle \nabla f(x_n), p_n \rangle \longrightarrow 0$.

Search routine

start: *if* $f(x_n + hp_n) < f(x_n)$ *then go to* first;

reduce: $h \leftarrow \frac{h}{2}$;

if $f(x_n + hp_n) \geq f(x_n)$ *then go to* reduce;

if $f\left(x_n + \frac{h}{2}p_n\right) \geq f(x_n + hp_n)$ *then* EXIT FROM ROUTINE NOW WITH $t_n = h$;

if f IS CONVEX *then* EXIT FROM ROUTINE NOW WITH $t_n = \frac{h}{2}$;

$h \leftarrow \frac{h}{2}$;

loop: *while* $f\left(x_n + \frac{h}{2}p_n\right) < f(x_n + hp_n)$ *do* $h \leftarrow \frac{h}{2}$;

EXIT FROM ROUTINE NOW WITH $t_n = h$;

first: *if* $x_n + 2hp_n$ IS IN C *then go to* inside;

$h \leftarrow \frac{h}{2}$;

go to start;

inside: *if* $f(x_n + 2hp_n) \geq f(x_n + hp_n)$ *then go to* oldway;

$t \leftarrow 2h$;

change: *while* $f[x_n + (t + h)p_n] < f(x_n + tp_n)$ *and* $x_n + (t + h)p_n$ IS IN C *do* $t \leftarrow t + h$;
EXIT FROM ROUTINE NOW WITH $t_n = t$;

oldway: *if* f IS CONVEX *and* $f(x_n + 2hp_n) \leq f(x_n)$ *then* EXIT FROM ROUTINE NOW WITH $t_n = h$;
go to loop;

4.10. DIRECTION ALGORITHMS FOR CONSTRAINED PROBLEMS

As we have mentioned before, since the step-size algorithms above guarantee that $\langle \nabla f(x_n), p_n \rangle \longrightarrow 0$, we must have a direction sequence such that this is a useful condition. For unconstrained problems—for example, with $p_n = p_n(x_n) = -\nabla f(x_n)$—the condition was $\| \nabla f(x_n) \| \longrightarrow 0$; under sufficient regularity assumptions, this implied that limit points x' of $\{x_n\}$ satisfied $\nabla f(x') = 0$ and that—say, for convex functionals—$\{x_n\}$ was a criticizing sequence. For constrained problems, analogously one should pick directions $p_n(x_n)$ so that in the limit the condition $\langle p(x), \nabla f(x) \rangle = 0$ is necessary for optimality and sufficient if f is convex. It is important here for this purpose that the directions p_n be feasible in our sense—i.e., $p_n \equiv x'_n - x_n$ for points x'_n and x_n in C.

EXERCISE. In $\mathbb{R}^l$, consider minimizing $f(x) = f(\xi_1, \ldots, \xi_l)$ over the set of x satisfying $x \geq 0$. For each x let direction $d(x) = (d_1, \ldots, d_l)$ have components

$$d_i = \begin{cases} 0 & \text{if } \xi_i = 0 \text{ and } \dfrac{\partial f}{\partial \xi_i} \geq 0 \\ -\dfrac{\partial f}{\partial \xi_i} & \text{otherwise} \end{cases}$$

and let $p(x) \equiv \alpha(x)\, d(x)$ for some scalar $\alpha(x)$ chosen so that p is feasible. Show that, if g is convex and $\langle \nabla f(x^*), d(x^*) \rangle = 0$, then x^* minimizes f. Sketch an example to show that choosing x_{n+1} to minimize $f[x_n + tp(x_n)]$ need not yield convergence to an optimal point. [The difficulty here is that $\langle d(x_n), \nabla f(x_n) \rangle \to 0$ *is* an optimality condition but $\langle p(x_n), \nabla f(x_n) \to 0$ is *not*.]

We shall now consider, for illustrative purposes, three methods of choosing directions $p_n(x_n)$ so that $\langle \nabla f(x_n), p_n(x_n) \rangle \longrightarrow 0$ is a useful condition. These methods can be combined with the step-size algorithms of Section 4.9 to yield complete numerical-minimization algorithms; we shall not state theorems concerning the resulting combined algorithms, although the reader should pause to consider such statements himself.

Recall that C is a convex set. Then a well-known necessary condition for x^* to minimize f over C, one that is sufficient if f is convex, is that $\langle x - x^*, \nabla f(x^*)\rangle \geq 0$ for all x in C—that is, every direction into C is a direction of increase for f. If one has a point x_n which does not satisfy this condition, then it is reasonable to seek the x'_n which most violates this condition and then take $p_n = x'_n - x_n$; this *conditional-gradient method* is well known [Demyanov–Rubinov (1967), Frank–Wolfe (1956), Gilbert (1966), Goldstein (1964a), Levitin–Poljak (1966a)]. Thus we seek x'_n such that

$$\langle \nabla f(x_n), x'_n - x_n\rangle \leq \inf_{x \in C} \langle \nabla f(x_n), x - x_n\rangle + \epsilon_n$$

for some nonnegative ϵ_n tending to zero. If C is bounded, we can always find x'_n; if C is bounded and norm-closed as well as convex, then we can take $\epsilon_n = 0$ if desired, although this causes unnecessary computation.

THEOREM 4.10.1. Let f be convex, bounded below on the bounded convex set C, and attain its minimum at some point x^* in C. Let x_n be a sequence in C such that $\langle \nabla f(x_n), p_n\rangle$ tends to zero, where $p_n \equiv x'_n - x_n$ and x'_n satisfies $\langle \nabla f(x_n), x'_n - x_n\rangle \leq \inf_{x \in C} \langle \nabla f(x_n), x - x_n\rangle + \epsilon_n$ for a sequence of nonnegative $\epsilon_n \longrightarrow 0$. Then $\{x_n\}$ is a minimizing sequence—that is, $f(x_n) \longrightarrow f(x^*)$.

Proof: From the convexity of f and the definition of x'_n we can write

$$\begin{aligned} 0 \leq f(x_n) - f(x^*) &\leq \langle \nabla f(x_n), x_n - x^*\rangle \\ &\leq \langle \nabla f(x_n), x_n - x'_n\rangle + \langle \nabla f(x_n), x'_n - x_n\rangle \\ &\quad - \langle \nabla f(x_n), x^* - x_n\rangle \\ &\leq -\langle \nabla f(x_n), p_n\rangle + \epsilon_n \end{aligned}$$

which tends to zero. Q.E.D.

EXERCISE. State some convergence theorems combining some step-size algorithms with the above direction algorithm.

The steepest-descent method for unconstrained problems, in which $p_n = -\nabla f(x_n)$, has been a popular method for many years, for some applications undeservedly. For constrained problems, that direction need not point into the constraint set C, so it is not directly applicable. Perhaps the most successful way of handling this has been to "project" the direction onto C; more precisely, one proceeds in the direction $p_n = x'_n - x_n$ where x'_n is the orthogonal projection onto C of $x_n - \alpha_n \nabla f(x_n)$ for some scalar $\alpha_n > 0$. This is the well-known *gradient-projection method* [Rosen (1960–61)]. In view of the numerical evidence that certain so-called *variable-metric methods* are much better than steepest descent for unconstrained problems [Chapter 7 of this volume, Fletcher–Powell (1963)] and the growing interest in such methods for constrained problems [Goldfarb (1966, 1969a, 1969b), Goldfarb–Lapidus

(1968)], we consider an analogous *variable-metric projected-gradient method.* We suppose that $\{A_n\}$ is a uniformly bounded, uniformly positive-definite family of self-adjoint linear operators on the space E—that is, that there are $m > 0$, $M < \infty$ such that $m\langle x, x\rangle \leq \langle A_n x, x\rangle \leq M\langle x, x\rangle$ for all x in E. For each n, let x'_n be the projection, with respect to the variable metric $\langle \cdot, A_n \cdot\rangle$, of $x_n - \alpha_n A_n^{-1}\nabla f(x_n)$ onto C; that is, x'_n minimizes

$$\langle x - [x_n - \alpha_n A_n^{-1}\nabla f(x_n)], A_n\{x - [x_n - \alpha_n A_n^{-1}\nabla f(x_n)]\}\rangle$$

over x in C. If C is norm-closed and convex, a unique x'_n exists. By the usual necessary condition, the variational definition of x'_n means that for all x in C we must have

$$\langle x - x'_n, A_n(x'_n - w_n)\rangle \geq 0 \tag{4.10.1}$$

where $w_n \equiv x_n - \alpha_n A_n^{-1}\nabla f(x_n)$. If we set $x \equiv x_n$ in this inequality, we obtain

$$\begin{aligned} 0 &\geq \langle x_n - x'_n, A_n(w_n - x'_n)\rangle \\ &= \langle x_n - x'_n, A_n(w_n - x_n)\rangle + \langle x_n - x'_n, A_n(x_n - x'_n)\rangle \end{aligned}$$

and since $w_n - x_n = -\alpha_n A_n^{-1}\nabla f(x_n)$, we obtain

$$\langle x_n - x'_n, -\alpha_n \nabla f(x_n)\rangle \leq -\langle x_n - x'_n, A_n(x_n - x'_n)\rangle$$

or

$$\alpha_n\langle -\nabla f(x_n), p_n\rangle \geq \langle p_n, A_n p_n\rangle \tag{4.10.2}$$

Therefore, the direction sequence is feasible. We now show that the condition $\lim_{n\to\infty} \langle \nabla f(x_n), p_n\rangle = 0$ is useful.

THEOREM 4.10.2. Let f be convex, bounded below on the norm-closed, bounded convex set C, and attain its minimum over C at x^*. Let x_n be a sequence in C such that the projected-gradient directions p_n defined above satisfy $\lim_{n\to\infty} \langle \nabla f(x_n), p_n\rangle = 0$ and $\alpha_n \geq \epsilon > 0$. Then $\{x_n\}$ is a minimizing sequence—that is, $f(x_n) \to f(x^*)$.

Proof: We write

$$\begin{aligned} 0 \leq f(x_n) - f(x^*) &\leq \langle \nabla f(x_n), x_n - x^*\rangle \\ &\leq \langle \nabla f(x_n), x_n - x'_n\rangle + \langle \nabla f(x_n), x'_n - x^*\rangle \\ &\leq \langle -\nabla f(x_n), p_n\rangle \\ &\qquad + \frac{1}{\alpha_n}\langle x_n - \alpha_n A_n^{-1}\nabla f(x_n) - x'_n, A_n(x^* - x'_n)\rangle \\ &\qquad + \frac{1}{\alpha_n}\langle x_n - x'_n, A_n(x'_n - x^*)\rangle \\ &\leq \langle -\nabla f(x_n), p_n\rangle + \frac{1}{\alpha_n}\langle x_n - x'_n, A_n(x'_n - x^*)\rangle \end{aligned}$$

by Equation 4.10.1. Therefore,

$$0 \leq f(x_n) - f(x^*) \leq \langle -\nabla f(x_n), p_n \rangle + \frac{1}{\alpha_n} M \| p_n \| \| x'_n - x^* \|$$
$$\leq \langle -\nabla f(x_n), p_n \rangle + \frac{1}{\alpha_n} M \| x'_n - x^* \| \left[\frac{\alpha_n}{m} \langle -\nabla f(x_n), p_n \rangle \right]^{1/2}$$

using Equation 4.10.2 and the positive-definiteness of A_n. Thus

$$0 \leq f(x_n) - f(x^*) \leq \langle -\nabla f(x_n), p_n \rangle + \frac{M \| x'_n - x^* \|}{\epsilon^{1/2} m^{1/2}} \langle -\nabla f(x_n), p_n \rangle^{1/2}$$

which tends to zero. Q.E.D.

EXERCISE. State some convergence theorems combining some step-size algorithms with the above direction algorithm.

EXERCISE. Show that, if $f(x) = \langle x^* - x, A(x^* - x) \rangle$, $A_n = A$, and $\alpha_n = 1$, then $x'_n = x^*$.

We note that our projected-gradient method for $A_n = I$, $E = \mathbb{R}^l$, and C a polyhedral set is not quite the same as the gradient-projection method originally described in Rosen (1960–61), since the latter requires that x'_n be the projection onto one of the faces to which x_n belongs or, in some implementations [Cross (1968)], onto a small neighborhood of x_n in C. The computational versions of gradient projection in use apply a special technique near edges of C which turns out to be essentially equivalent to bounding α_n away from zero but keeping it small enough so that the projection is always very near x_n. Thus it is clear that a simple convergence proof for Rosen's original computational gradient-projection method can be fashioned in this way from our results above; this has been done [Kreuser (1969)]. If one, however, does not take α_n small, one needs a good, efficient method for projection, in an arbitrary quadratic metric, onto a full polyhedral set. Such an algorithm has been brought to our attention [Golub-Saunders (1969)] and raises the possibility of using larger α_n, which may well be more powerful than the original gradient-projection approach, at least far away from the solution.

The method analyzed in Theorem 4.10.1 can be considered intuitively in a fashion different from that presented before if we notice that x'_n is chosen so as to (approximately) minimize $f(x_n) + \langle \nabla f(x_n), x - x_n \rangle$ over C; that is, x'_n solves the problem with the cost function *linearized* at x_n. We consider next the approximation of f by a *quadratic* at x_n (*Newton's method*)—namely,

$$f(x_n) + \langle \nabla f(x_n), x - x_n \rangle + \tfrac{1}{2} \langle f''_n(x - x_n), x - x_n \rangle$$

where f''_n denotes the derivative f''_{x_n}. If x'_n is chosen to minimize

$$g_n(x) \equiv \langle \nabla f(x_n), x - x_n \rangle + \tfrac{1}{2} \langle f''_n(x - x_n), x - x_n \rangle$$

then

$$\langle \nabla g(x'_n), x - x'_n \rangle \equiv \langle \nabla f(x_n) + f''_n(x'_n - x_n), x - x'_n \rangle \geq 0$$

for all x in C. Setting $x = x_n$ and defining $p_n = x'_n - x_n$ in this inequality yields

$$\langle \nabla f(x_n), p_n \rangle \leq -\langle p_n, f''_n p_n \rangle \leq 0$$

for convex f, implying that p_n is a direction of nonincreasing f-values as needed.

THEOREM 4.10.3. Let f be convex, bounded below on the norm-closed, bounded convex set C, and attain its minimum over C at x^*; let f''_x exist in C and $\| f''_x \| \leq B > 0$ for all x in C. Let $\{x_n\}$ be a sequence in C such that $\lim_{n \to \infty} \langle \nabla f(x_n), p_n \rangle = 0$ where $p_n = x'_n - x_n$ and x'_n minimizes

$$g_n(x) \equiv \langle \nabla f(x_n), x - x_n \rangle + \tfrac{1}{2} \langle f''_n(x - x_n), x - x_n \rangle$$

over C. Then $f(x_n) \longrightarrow f(x^*)$.

Proof: By the definition of x'_n as we saw above, we have

$$\langle -\nabla f(x_n), p_n \rangle \geq \langle f''_n p_n, p_n \rangle$$

However,

$$\langle f''_n p_n, p_n \rangle \geq \frac{1}{B} \| f''_n p_n \|^2$$

and hence from $\langle -\nabla f(x_n), p_n \rangle \longrightarrow 0$ we conclude that $\| f''_n p_n \| \longrightarrow 0$. Thus for all x in C, we write

$$\begin{aligned} &\langle x'_n - x_n, \nabla f(x_n) \rangle - \langle x - x_n, \nabla f(x_n) \rangle \\ &\quad = \langle x'_n - x, \nabla f(x_n) + f''_n(x'_n - x_n) \rangle - \langle x'_n - x, f''_n(x'_n - x_n) \rangle \\ &\quad = -\langle x - x'_n, \nabla g_n(x'_n) \rangle - \langle x'_n - x, f''_n p_n \rangle \leq -\langle x'_n - x, f''_n p_n \rangle \end{aligned}$$

since $\langle \nabla g_n(x'_n), x - x'_n \rangle \geq 0$. By the boundedness of C and the fact that $\| f''_n p_n \| \longrightarrow 0$, the numbers

$$\epsilon_n \equiv \sup_{x \in C} | \langle x'_n - x, f''_n p_n \rangle |$$

tend to zero. Thus we have

$$\langle x'_n - x_n, \nabla f(x_n) \rangle \leq \langle x - x_n, \nabla f(x) \rangle + \epsilon_n$$

for all x in C with $\epsilon_n \longrightarrow 0$. The result now follows from Theorem 4.10.1. Q.E.D.

EXERCISE. State some convergence theorems combining some step-size algorithms with the above direction algorithm.

A local convergence theorem yielding the usual quadratic convergence rate when $x_{n+1} = x'_n$ has been given in Levitin–Poljak (1966a).

EXERCISE. Contrary to the unconstrained case (see Section 7.2), by considering the minimization of $x^2 + y^2$ over $\{(x, y); y \geq 1\}$, show that picking x_{n+1} so that $f(x_{n+1}) \leq f(x'_n)$ *need not* maintain quadratic convergence, where x'_n is generated by the above Newton's method.

For more extensive discussions of algorithms for constrained problems, the reader is referred to Fiacco–McCormick (1968), Zangwill (1969), and references therein.

4.11. OTHER METHODS FOR CONSTRAINED PROBLEMS

A considerably different kind of method has been developed for the case in which the constraints take the form $P(x) = 0$ where $P: E \longrightarrow E$ is nonlinear, implying that one need not be able to proceed from x_n into C along straight lines. In this case, under suitable hypotheses [Altman (1966b)], one can find $s(x, t) \in E$ such that $P[x - t\nabla f(x) + s(x, t)] = 0$ for all $t \geq 0$ and $\|s(x, t)\| \leq Kt^2$ for some K. Thus only a small perturbation of the linear motion keeps us in C. Algorithms have been given for determining t-values, and convergence proofs are known. The methods for computing $s(x, t)$, however, are very complex and do not appear to lend themselves to practical computation; therefore, we consider the method no further.

One further type of method for constrained problems which we wish to consider is the *penalty-function method.* We have met this approach before in Sections 3.2 and 3.3 in a more specialized form. In fact, the whole approach fits into the discretization analysis if one makes some extensions in those results, but this adds but little to the general applicability of those theorems; therefore, we treat the penalty-function method briefly in the more classical fashion.

We seek to minimize $f(x)$ over $C = \{x; g(x) \leq 0\}$, where g is some nonlinear functional. Instead, we shall approximately minimize $f(x) + P_n[g(x)]$ over E, where the *penalty functions* P_n are such that, for $t > 0$,

$$\lim_{n \to \infty} P_n(t) = \infty, \qquad \text{uniformly for} \quad t \geq \delta > 0 \quad \text{for all} \quad \delta > 0$$

Thus P_n will penalize us for having an x with $g(x) > 0$.

EXERCISE. Give some examples of penalty functions that satisfy the conditions immediately above.

What we can hope will occur, then, is that our computed sequence x_n will satisfy

$$\limsup_{n\to\infty} g(x_n) \leq 0$$

This, however, is not enough in general to guarantee that

$$d(x_n, C) = \inf_{x \in C} ||x_n - x||$$

is tending to zero.

DEFINITION 4.11.1 [Levitin–Poljak (1966a)]. The constraint defined by g is called *correct* if $\limsup_{n\to\infty} g(x_n) \leq 0$ implies $\lim_{n\to\infty} d(x_n, C) = 0$.

> **EXERCISE.** Find some explicit conditions under which constraints are correct.

THEOREM 4.11.1. Let g define a correct constraint; for some $\epsilon > 0$ let $|f(x) - f(y)| \leq L||x - y||$ if $d(x, C) \leq \epsilon$ and $d(y, C) \leq \epsilon$; let $P_n[g(x)] \geq 0$ for all $x \in E$; let $\lim_{n\to\infty} P_n(t) = \infty$ for $t > 0$, uniformly for $t \geq \delta > 0$ for all $\delta > 0$; and let $\lim_{n\to\infty} P_n[g(x)] = 0$ for all $x \in C_0$, a dense subset of C. Define

$$m_n = \inf_{x \in E} \{f(x) + P_n[g(x)]\}, \qquad m = \inf_{x \in C} f(x)$$

and assume $\inf_{x \in E} f(x) = \bar{m} > -\infty$. For a sequence $\epsilon_n > 0$, $\epsilon_n \to 0$, let $x_n \in E$ satisfy

$$f(x_n) + P_n[g(x_n)] \leq m_n + \epsilon_n$$

Then $\{x_n\}$ is an approximate minimizing sequence for f over C in the sense of Definition 1.6.1.

Proof: Let $w_n \in C$, $\lim_{n\to\infty} f(w_n) = m$. Since f is continuous, there exists $w'_n \in C_0$ with

$$|f(w'_n) - f(w_n)| \leq f(w_n) - m + \epsilon_n$$

Thus $\lim_{n\to\infty} f(w'_n) = m$ also. For each integer r,

$$f(x_n) + P_n[g(x_n)] \leq f(w'_r) + P_n[g(w'_r)] + \epsilon_n$$

By picking r large and then n large, we conclude that

$$\limsup_{n\to\infty} \{f(x_n) + P_n[g(x_n)]\} \leq m$$

Since $P_n[g(x_n)] \geq 0$, we also have $\limsup_{n\to\infty} f(x_n) \leq m$. Also

$$P_n[g(x_n)] = \{f(x_n) + P_n[g(x_n)]\} - f(x_n)$$

which implies

$$\limsup_{n\to\infty} P_n[g(x_n)] \leq m - \bar{m} < \infty$$

Therefore, also

$$\limsup_{n\to\infty} g(x_n) \leq 0$$

or otherwise for some subsequence x_{n_i} we would have $P_{n_i}[g(x_{n_i})] \longrightarrow \infty$, a contradiction. Since g is a correct constraint, $d(x_n, C) \longrightarrow 0$. Thus, for large n, $d(x_n, C) < \epsilon$ and we write

$$|f(x_n) - f(x'_n)| \leq L\|x_n - x'_n\| \leq 2Ld(x_n, C)$$

where

$$x'_n \in C, \qquad \|x_n - x'_n\| \leq 2d(x_n, C)$$

Then we have

$$m \leq f(x'_n) = f(x_n) + f(x'_n) - f(x_n) \leq f(x_n) + 2Ld(x_n, C)$$

so that

$$f(x_n) \geq m - 2Ld(x_n, C)$$

which implies

$$\liminf_{n\to\infty} f(x_n) \geq m$$

We already have

$$\limsup_{n\to\infty} f(x_n) \leq m$$

so, therefore, $\{x_n\}$ is an approximate minimizing sequence. Q.E.D.

As usual, one can use the results of Section 1.6 to deduce stronger convergence results.

It should be pointed out that the problem of minimizing a convex function, bounded below, over a bounded convex set can be reduced to that of a linear function over a convex set. For example, suppose that C is a convex bounded set in E and that f is a convex functional. Define the new space $E_1 = E \times \mathbb{R}$, $f_1(x, \lambda) = \lambda$, and for some $\bar{x} \in C$ let

$$C_1 = \{(x, \lambda); x \in C, f(x) \leq \lambda \leq f(\bar{x})\}$$

The original problem is solved now by minimizing the linear functional f_1

over the convex bounded set C_1. Since $\nabla f_1 = (0, 1)$, it is Lipschitz-continuous with constant L arbitrarily small, allowing for application of the preceding methods, in particular, gradient projection.

Finally, we mention briefly another method, somewhat similar to the penalty-function method, but for which the penalty is introduced in a different way. For simplicity only, consider the problem of minimizing a strictly convex functional f such that $f(x) \longrightarrow +\infty$ as $||x|| \longrightarrow \infty$ over the convex subset C of $\mathbb{R}^l$ defined by

$$C = \{x; g_i(x) \leq 0, i = 1, \ldots, k\}$$

where the g_i are convex functionals; we suppose C has an interior point. The *method of centers* proceeds as follows [Céa (1969), Huard (1967), Zangwill (1969)]; given an initial $x_0 \in C$, we compute x_{n+1} as a point minimizing

$$f_n(x) \equiv [f(x) - f(x_n)] \prod_{i=1}^{k} [-g_i(x)] \quad \text{over} \quad \{x; f(x) \leq f(x_n)\}$$

If an iterative method is used to locate x_{n+1} starting with x_n as a first approximation, since $f_n(x_n) = 0$, once an iterate z with $f_n(z) < 0, f(z) < f(x_n)$ is found, gradient methods can never increase the f_n-values, and hence $\prod_{i=1}^{k} [-g_i(x)]$ will always be positive and $f(x) - f(x_n)$ will always be negative; therefore, the iterative method to compute x_{n+1} can ignore the constraints, just as in the penalty-function method. Under the hypotheses we have stated, it is known that $||x_n - x^*|| \longrightarrow 0$, where x^* is the solution to the minimization problem. Similar results hold for much more general versions of this method.

We remark again that there are many other methods for constrained minimization; since this is not a book on mathematical-programming methods, we have only mentioned a few.

General reference: Levitin–Poljak (1966a).

5 CONJUGATE-GRADIENT AND SIMILAR METHODS IN HILBERT SPACE

5.1. INTRODUCTION

In recent years there has been a great deal of interest in iterative minimization methods, for both constrained and unconstrained problems, which make use of the idea of conjugate directions; we shall discuss some of the practical algorithms in $\mathbb{R}^l$ in other chapters. In the present chapter we wish to describe, in a general setting, the basic theory behind conjugate-direction and particularly conjugate-gradient methods. We shall examine the method first for the simple case in which the function f to be minimized without constraints is a quadratic functional, essentially with $0 < aI \leq f''_x \leq AI$. It is the great power of the methods when applied to this problem that has made them appear attractive for the more general nonlinear problems. Later we shall extend the results to the more general case. We shall throughout this chapter consider the problem as defined over a real, separable Hilbert space $\mathscr{H}$ with inner product $\langle \cdot, \cdot \rangle$.

5.2. CONJUGATE DIRECTIONS FOR QUADRATIC FUNCTIONALS

If we seek a critical point of a quadratic functional f, then we are really trying to solve a linear equation. Thus let M be a bounded linear operator with bounded inverse from $\mathscr{H}$ into $\mathscr{H}$. Let H be a positive-definite, self-adjoint, bounded linear operator from $\mathscr{H}$ into $\mathscr{H}$; then $N \equiv M^*HM$ has the same properties. The problem of solving $Mx = k$ for given k with $h = M^{-1}k$ now can be stated as the problem of minimizing the functional $f(x) = \langle r, Hr \rangle$

where $r = k - Mx$ is the residual. When we consider more general functionals f, we shall still find it necessary to examine a functional $\langle r, Hr \rangle$, so for these later uses we shall define

$$E(x) \equiv \langle r, Hr \rangle = \langle k - Mx, H(k - Mx) \rangle = \langle h - x, N(h - x) \rangle$$

Since N is positive-definite and $E''_x = 2N$, we see that E is strictly minimized over $\mathscr{H}$ at $x = h$. We shall attempt to minimize E over $\mathscr{H}$ by minimizing it over a sequence of expanding subspaces.

THEOREM 5.2.1. Let $\{B_n\}$ be a sequence of closed linear subspaces of $\mathscr{H}$, $B_n \subset B_{n+1}$, $B \equiv$ norm closure of $\bigcup_{n=1}^{\infty} B_n$. Let x_n minimize E on B_n. Then $x_n \longrightarrow x' \in B$, and x' minimizes E over B.

Proof: The points x_n exist and are unique because of the growth property and the uniform (quasi-)convexity of $E(x)$. Since $E(x_n)$ forms a decreasing sequence bounded below by zero, $\{E(x_n)\}$ is a Cauchy sequence. For $n \geq m$, we write

$$\langle x_m - x_n, N(x_m - x_n) \rangle = E(x_m) - E(x_n) + 2\langle M^*Hr_n, x_m - x_n \rangle$$

Since $n \geq m$, $x_m - x_n$ is in B_n; since x_n minimizes $E(x)$ over B_n, $\nabla E(x_n)$, which equals $-2M^*Hr_n$, is orthogonal to B_n and therefore to $x_m - x_n$. Thus we conclude that

$$\langle x_m - x_n, N(x_m - x_n) \rangle = E(x_m) - E(x_n)$$

which tends to zero, since $\{E(x_i)\}$ is a Cauchy sequence. Since N is positive-definite, $\{x_i\}$ is a Cauchy sequence and there exists $x' \in B$ such that $x_n \longrightarrow x'$. By a continuity argument we find, setting $r' = k - Mx'$, that $\langle M^*Hr', z \rangle = 0$ for all $z \in B$. Since again $\nabla E(x') = -2M^*Hr'$, this implies that x' minimizes E over B. Q.E.D.

EXERCISE. Prove that $\langle M^*Hr', z \rangle = 0$ for all $z \in B$, where r' is defined in the *Proof* of Theorem 5.2.1 above.

As a practical matter, the minimization is easier if B_n is finite-dimensional—if, say, B_n is spanned by the linearly independent vectors $\{p_0, p_1, \ldots, p_{n-1}\}$ for all n. Then of course $x_n = \sum_{j=0}^{n-1} \alpha_{n,j} p_j$. It would be convenient if $\alpha_{n,j}$ were independent of n, so that $x_{n+1} = x_n + \alpha_n p_n$. It is a simple matter to show that this is the case if and only if either $x_n = x_{n+1}$ or $\langle p_i, Np_j \rangle = 0$ for $i = 0, 1, \ldots, j - 1$ [Antosiewicz–Rheinboldt (1962)]. We include the proof in one direction as a part of the following.

THEOREM 5.2.2. Let $\{p_n\}_0^\infty$ be a sequence of linearly independent vectors satisfying $\langle p_i, Np_j\rangle = 0$ if $i \neq j$ and let x_0 be arbitrary. Let

$$x_{n+1} = x_n + c_n p_n, \qquad c_n = \frac{\langle M^*Hr_n, p_n\rangle}{\langle p_n, Np_n\rangle}, \qquad r_n \equiv k - Mx_n$$

Let B_n be spanned by $p_0, \ldots, p_{n-1}$ and let $B =$ closure of $\bigcup_1^\infty B_n$. Then $x_n \to x'$ minimizing E over B.

Proof: For $i \leq n - 1$,

$$\langle M^*Hr_n, p_i\rangle = \langle M^*Hr_{n-1}, p_i\rangle - c_{n-1}\langle Np_{n-1}, p_i\rangle$$

For $i < n - 1$ we have

$$\langle M^*Hr_n, p_i\rangle = \langle M^*Hr_{n-1}, p_i\rangle$$

while, by the definition of c_i, $\langle M^*Hr_{i+1}, p_i\rangle = 0$. Thus $\langle M^*Hr_n, z\rangle = 0$ for all z in B_n and hence x_n minimizes E over B_n. The rest follows from Theorem 5.2.1. Q.E.D.

EXERCISE. Prove the converse of Theorem 5.2.2 by proving that the $\alpha_{n,j}$ defined immediately before Theorem 5.2.2 are independent of n only if the directions $\{p_n\}_0^\infty$ satisfy $\langle p_i, Np_j\rangle = 0$ for $i \neq j$.

A set of linearly independent directions $\{p_n\}_0^\infty$ satisfying $\langle p_i, Np_j\rangle = 0$ if $i \neq j$ is called a set of *N-conjugate* (or *conjugate*) *directions*. A general scheme has been devised [Hestenes (1956)] by which such directions can be generated. One can show fairly directly that the following is valid [Daniel (1965, 1967b), Hestenes (1956)].

PROPOSITION 5.2.1. Let K, N be positive-definite, bounded, self-adjoint linear operators in $\mathscr{H}$, and let $g_0 \neq 0$ be given in $\mathscr{H}$. The algorithm

$$g_{n+1} = g_n - c_n Np_n, \qquad p_0 = Kg_0, \qquad p_{n+1} = Kg_{n+1} + b_n p_n$$

with

$$c_n = \frac{\langle g_n, p_n\rangle}{\langle p_n, Np_n\rangle} \quad \text{and} \quad b_n = \frac{-\langle Np_n, Kg_{n+1}\rangle}{\langle p_n, Np_n\rangle}$$

generates directions satisfying $\langle g_i, Kg_j\rangle = \langle p_i, Np_j\rangle = 0$ if $i \neq j$, $\langle g_{i+1}, p_i\rangle = 0$, $\langle g_i, p_i\rangle = \langle g_i, Kg_i\rangle$,

$$b_i = \frac{\langle g_{i+1}, Kg_{i+1}\rangle}{\langle g_i, Kg_i\rangle}$$

The algorithm terminates at $n = n_0$ if and only if $g_{n_0} = 0$. If we define

$$\mu(x) = \frac{\langle x, Nx \rangle}{\langle x, K^{-1}x \rangle}, \qquad \nu(x) = \frac{\langle x, Kx \rangle}{\langle x, N^{-1}x \rangle}, \qquad T = KN$$

then the spectrum of T lies in an interval $[a, A]$, $a > 0$, and for any such a, A, we have

$$a \leq \mu(p_i) \leq \frac{1}{c_i} \leq \mu(Kg_i) \leq A \quad \text{and} \quad a \leq \nu(g_i) \leq \frac{1}{c_i} \leq \nu(Np_i) \leq A$$

According to Theorem 5.2.2, the iteration defined therein yields a solution to $Mx = k$ if $B = \mathscr{H}$; if $B \neq \mathscr{H}$, x' need not equal h in general. Of course, if $\mathscr{H}$ is finite-dimensional, the iteration terminates, $B = \mathscr{H}$, and $x' = h$. For infinite-dimensional problems, however, we need additional conditions to assure $x' = h$.

EXERCISE. Find an example of a conjugate-direction method for a specific problem for which the limit $x' \neq h$.

General references: Hayes (1954), Hestenes–Stiefel (1952).

5.3. CONJUGATE GRADIENTS FOR QUADRATIC FUNCTIONALS

We consider a special conjugate-direction algorithm—namely, one in which, in the algorithm of Proposition 5.2.1, we take $g_0 = M^*Hr_0 = -\frac{1}{2}\nabla E(x_0)$. Clearly, then, $g_n = M^*Hr_n$, which implies that the c_n of Proposition 5.2.1 and Theorem 5.2.2 are the same if we write $x_{n+1} = x_n + c_n p_n$. Thus $x_n \longrightarrow x'$ minimizing E on some closed subspace of $\mathscr{H}$. If $K = I$, then $g_n = M^*Hr_n = -\frac{1}{2}\nabla E(x_n)$ and, since $p_{n+1} = Kg_{n+1} + b_n p_n$, we see that the new direction p_{n+1} is obtained by "conjugatizing" the direction $-\frac{1}{2}\nabla E(x_n)$—that is, by projecting $-\frac{1}{2}\nabla E(x_n)$ onto the space of vectors conjugate to p_0, $p_1, \ldots, p_n$; hence the name *conjugate-gradient* method. We shall return later to the projection aspect of the method. We now wish to show that, for the conjugate-gradient method, $x_n \longrightarrow h$.

THEOREM 5.3.1. Using the conjugate-gradient method, $x_n \longrightarrow h$. We have

$$E(x_{n+1}) \leq E(x_n) \cdot q, \qquad 0 \leq q < 1$$

Let a, A be the positive spectral bounds for $T \equiv KN$. If $KN = NK$, then we can take

$$q = \left(\frac{A - a}{A + a}\right)^2$$

Otherwise, we can take $q = 1 - (a/A)$. The same convergence rates obtain for $\|x_n - h\|^2$.

Proof: It is trivial to verify that

$$E(x_n) - E(x_{n+1}) = c_n\langle g_n, Kg_n\rangle = c_n^2\langle p_n, Np_n\rangle$$

Next we have

$$\begin{aligned} E(x_i) &= \langle r_i, Hr_i\rangle = \langle M^{-1}r_i, M^*Hr_i\rangle = \langle M^{-1}(M^*H)^{-1}M^*Hr_i, M^*Hr_i\rangle \\ &= \langle N^{-1}g_i, g_i\rangle = \frac{1}{\nu(g_i)}\langle g_i, Kg_i\rangle \end{aligned}$$

$$E(x_i) - E(x_{i+1}) = c_i\langle g_i, Kg_i\rangle = E(x_i)c_i\nu(g_i)$$

The estimates of the theorem follow from $c_i \geq 1/A$, $\nu(g_i) \geq a$. If K and N commute, then

$$a_i\nu(g_i) \geq \frac{\nu(g_i)}{\mu(Kg_i)} = \frac{[Kg_i, Kg_i]^2}{[Kg_i, TKg_i][Kg_i, T^{-1}Kg_i]}$$

where $[x, y] \equiv \langle x, K^{-1}y\rangle$. It is easy to see that T is self-adjoint positive-definite relative to $[\cdot, \cdot]$ with spectral bounds a, A; thus

$$c_i\nu(g_i) \geq \frac{4aA}{(A + a)^2}$$

by the inequality of Kantorovich [Faddeev–Faddeeva (1963), Kantorovich (1948)]. Now let $\beta > 0$ be the lower spectral bound for N. Then

$$\beta\,\|x_n - h\|^2 \leq \langle h - x_n, N(h - x_n)\rangle = E(x_n) \leq q^n E(x_0)$$

Thus

$$\|x_n - h\|^2 \leq q^n\frac{E(x_0)}{\beta}, \qquad x_n \longrightarrow h$$

and the stated convergence rate is valid. Q.E.D.

It is also possible to show that another error measure—namely,

$$F(x) \equiv \langle h - x, K^{-1}(h - x)\rangle$$

decreases [Hestenes–Stiefel (1952), Daniel (1965, 1967b)].

EXERCISE. Prove that

$$F(x_n) - F(x_{n+1}) \geq \langle x_{n+1} - x_n, K^{-1}(x_{n+1} - x_n)\rangle$$

In some cases the method can be shown to converge even when $a = 0$, but examples are known in which we then have $\| x_n - x^* \| \geq (\ln n)^{-\lambda}$ for some $\lambda > 0$, showing that no geometric convergence rate is possible [Odloleskal (1969), Poljak (1969a)].

General references: Antosiewicz–Rheinboldt (1962), Daniel (1965, 1967b), Hayes (1954), Hestenes (1956), Hestenes–Stiefel (1952).

5.4. CONJUGATE GRADIENTS AS AN OPTIMAL PROCESS

Much-improved bounds on the convergence rate can be obtained by viewing the conjugate-gradient method in a different light, one which shows more clearly the great power of the method—as opposed, say, to the steepest-descent method, which also has a convergence factor like $(A - a)/(A + a)$.

Suppose we seek to solve $Mx = k$—that is, $M^*HMx = M^*Hk$—by some sort of gradient method; for more generality we allow ourselves to multiply gradients also by an operator K, where M, H, N, K, T are as defined earlier. If at each step we allow ourselves to make use of all previous information, we are lead to consider iterations of the form

$$x_{n+1} = R_n(T)x_0 + P_n(T)Th, \qquad h = M^{-1}k$$

where $R_n(\lambda)$, $P_n(\lambda)$ are polynomials of degree less than or equal to n. If we should by chance have $x_0 = h$, we would want $x_n = h$ for all n. This leads, since h should be considered arbitrary, to the requirement that

$$x_{n+1} = x_0 + P_n(T)T(h - x_0) \tag{5.4.1}$$

where $P_n(\lambda)$ is a polynomial of degree less than or equal to n.

We wish to use methods of spectral analysis to discuss such methods, so we are forced to assume that

$$N = \rho(T)$$

where $\rho(\lambda)$ is a positive function continuous on some neighborhood of the spectrum of T. As we shall later see, this is satisfied in the practical methods, where usually $\rho(\lambda) \equiv \lambda$ or $\rho(\lambda) \equiv 1$. For each n, we wish to choose $P_n(\lambda)$ so that $E(x_{n+1})$ is the least possible under any method of the form of Equation 5.4.1. According to the spectral theorem, we can write

$$E(x_{n+1}) = \int_a^A \rho(\lambda)[1 - \lambda P_n(\lambda)]^2 \, ds(\lambda) \tag{5.4.2}$$

where $s(\lambda)$ is a known increasing function. The fact that there is a polynomial $P_n(\lambda)$ yielding this least value follows from a straightforward generalization

[Daniel (1965, 1967b)] of the theorem in finite dimensions as proved in Stiefel (1954, 1955).

PROPOSITION 5.4.1. The error measure $E(x_{n+1})$ is minimized by setting $R_{n+1}(\lambda) \equiv 1 - \lambda P_n(\lambda)$ to be the $(n + 1)$st element of the orthogonal {on $[a, A]$ relative to the weight function $\lambda p(\lambda)\, ds(\lambda)$} set of polynomials $R_i(\lambda)$ satisfying $R_i(0) = 1$.

EXERCISE. Prove Proposition 5.4.1.

We shall now show that, for each n, the vectors generated by the conjugate-gradient method are precisely those generated by this optimal process.

THEOREM 5.4.1. For each n, the vector x_n generated by the conjugate-gradient (CG) method coincides with that generated by the optimal process of the form in Equation 5.4.1.

Proof: Given n, the vectors $p_0, \ldots, p_{n-1}$ in the CG method are independent. Since $p_0 = Kg_0$ and $p_{i+1} = Kg_{i+1} + b_i p_i$, it is clear that any linear combination of $p_0, \ldots, p_{n-1}$ can be written as a linear combination of $Kg_0, \ldots, Kg_{n-1}$. Thus the n vectors $Kg_0, \ldots, Kg_{n-1}$ span at least the n-dimensional space $B_n \equiv sp\{p_0, \ldots, p_{n-1}\}$ and hence $B_n = sp\{Kg_0, \ldots, Kg_{n-1}\}$. Now $Kg_0 = T^0 Kg_0$; assume that for $j \leq i$, Kg_j can be written as a linear combination of $T^0 Kg_0, T^1 Kg_0, \ldots, T^j Kg_0$. Then

$$Kg_{i+1} = K(g_i - c_i N p_i) = Kg_i - c_i T p_i$$

We can write p_i as a linear combination of $Kg_0, \ldots, Kg_i$, each of which, by the inductive assumption, is a linear combination of $T^0 Kg_0, \ldots, T^i Kg_0$. Therefore, Kg_{i+1} is a linear combination of $T^0 Kg_0, \ldots, T^{i+1} Kg_0$. Reasoning as above, we have

$$B_n = sp\{T^0 Kg_0, \ldots, T^{n-1} Kg_0\}$$

Now x_n minimizes $E(x)$ on $x_0 + B_n$ if x_n is generated by the CG method. By what we have shown above, this says that the x_n generated by the CG method minimizes $E(x)$ on the set of points

$$x = x_0 + \sum_{i=0}^{n-1} s_i T^i K g_0 = x_0 + P_{n-1}(T)T(h - x_0)$$

where $P_{n-1}(\lambda)$ is the $(n - 1)$st-degree polynomial $P_{n-1}(\lambda) = \sum_{i=0}^{n-1} s_i \lambda^i$. That is, among all iterations of the form

$$x_{n+1} = x_0 + P_n(T)T(h - x_0)$$

the CG method makes $E(x_{n+1})$ the least. Q.E.D.

Thus, if we insert any polynomial into Equation 5.4.2, we can get a bound for $E(x_{n+1})$ where x_{n+1} is generated by the conjugate-gradient method, since that method gives the least value of $E(x_{n+1})$. If we choose for comparison $1 - \lambda P_n(\lambda)$ as the $(n + 1)$st Chebyshev polynomial relative to $\lambda\rho(\lambda)\, ds(\lambda)$ on $[a, A]$, we find the following bound.

PROPOSITION 5.4.2. Let $\alpha = a/A$. Then, for the conjugate-gradient method,

$$E(x_n) \leq w_n^2 E(x_0) \leq 4\left(\frac{1 - \sqrt{\alpha}}{1 + \sqrt{\alpha}}\right)^{2n} E(x_0)$$

and $\|x_n - h\|^2$ converges to zero at this same rate, where

$$w_n \equiv \frac{2(1 - \alpha)^n}{(1 + \sqrt{\alpha})^{2n} + (1 - \sqrt{\alpha})^{2n}}$$

EXERCISE. Prove Proposition 5.4.2.

By this result we have reduced our estimate of the convergence factor from $(1 - \alpha)/(1 + \alpha)$ to at least $(1 - \sqrt{\alpha})/(1 + \sqrt{\alpha})$. When one uses the steepest-descent algorithm to solve $Mx = k$ by minimizing E, one moves from x_n to x_{n+1} in the direction M^*Hr_n. Therefore, the steepest-descent method has the form of Equation 5.4.1 and, therefore, reduces the error $E(x)$ by *less* than the conjugate-gradient method for every n. Since the best-known and in certain cases best possible convergence estimates for steepest descent [Akaike (1959)] are of the form $(1 - \alpha)/(1 + \alpha)$, while we have at least $(1 - \sqrt{\alpha})/(1 + \sqrt{\alpha})$, we see that the convergence of the conjugate-gradient method is also asymptotically better.

For clarity, we now state the form that the conjugate-gradient algorithm takes in certain special cases.

The iteration takes its simplest form in the case in which the operator M is itself positive-definite and self-adjoint; it was this case for which the method was originally developed. Here we may now take $H = M^{-1}$ and $K = I$. Thus

$$N = T = M, \qquad E(x) = \langle h - x, M(h - x)\rangle$$

Since $N = T$, we have $\rho(\lambda) \equiv \lambda$, and the analysis of this section applies. The iteration becomes as follows:

Given x_0, let $p_0 = r_0 = k - Mx_0$. For $n = 0, 1, \ldots$, let

$$c_n = \frac{\|r_n\|^2}{\langle p_n, Mp_n\rangle} = \frac{\langle r_n, p_n\rangle}{\langle p_n, Mp_n\rangle}, \qquad x_{n+1} = x_n + c_n p_n$$

$$r_{n+1} = r_n - c_n M p_n, \qquad p_{n+1} = r_{n+1} + b_n p_n$$

where

$$b_n = -\frac{\langle r_{n+1}, Mp_n \rangle}{\langle p_n, Mp_n \rangle} = \frac{||r_{n+1}||^2}{||r_n||^2}$$

A second special case which is simple enough for practical use arises from setting $H = K = I$, so that $T = N = M^*M$. Again, $\rho(\lambda) \equiv \lambda$, and we have $E(x) = ||r||^2$. Fortunately, for computational purposes one can avoid the actual calculation of M^*M and can put the iteration in the following form:

Given x_0, let $r_0 = k - Mx_0, p_0 = g_0 = M^*r_0$. For $n = 0, 1, \ldots$, let

$$c_n = \frac{\langle g_n, p_n \rangle}{\langle Mp_n, Mp_n \rangle} = \frac{||g_n||^2}{||Mp_n||^2}, \qquad x_{n+1} = x_n + c_n p_n$$

$$r_{n+1} = r_n - c_n Mp_n, \qquad g_{n+1} = M^* r_{n+1}$$

$$p_{n+1} = g_{n+1} + b_n p_n$$

where

$$b_n = -\frac{\langle Mp_n, Mg_{n+1} \rangle}{||Mp_n||^2} = \frac{||g_{n+1}||^2}{||g_n||^2}$$

A third special case arises from $H = (M^*M)^{-1}$, $K = M^*M$, so that $N = I, T = M^*M$, $\rho(\lambda) \equiv 1$, $E(x) = ||h - x||^2$. By some manipulation, the iteration takes the following form:

Given x_0, let $r_0 = k - Mx_0, p_0 = M^*r_0$. For $n = 0, 1, \ldots$, let

$$c_n = \frac{||r_n||^2}{||p_n||^2}, \qquad x_{n+1} = x_n + c_n p_n$$

$$r_{n+1} = r_n - c_n Mp_n, \qquad p_{n+1} = M^* r_{n+1} + b_n p_n$$

where

$$b_n = \frac{||r_{n+1}||^2}{||r_n||^2}.$$

EXERCISE. Show that the last two algorithms above generate the desired iterates.

General references: Daniel (1965, 1967b), Faddeev–Faddeeva (1963).

5.5. THE PROJECTED-GRADIENT VIEWPOINT

It has been widely believed that the CG method exhibits superlinear convergence—that is, that $||x_n - h||$ tends to zero faster than any geometric sequence λ^n with $\lambda > 0$—although the best error estimates in general only yield

$$\lambda = \frac{\sqrt{A} - \sqrt{a}}{\sqrt{A} + \sqrt{a}}$$

If we view the method as one of projecting the gradient direction onto the space conjugate to all preceding directions, we obtain an indication that the convergence might in fact be superlinear; the result we obtain in this way is also needed later for the analysis of nonquadratic functionals. For simplicity of notation, we restrict ourselves to the simplest special case of the CG method with M itself positive-definite and self-adjoint, with $N = T = M$, $K = I$.

Without loss of generality, we consider the CG iteration starting with a first guess $x_0 = 0$. Suppose we are given a vector $d \neq 0$ such that $\langle d, k \rangle = 0$. We define an equivalent inner product $[\cdot, \cdot]$, by

$$[x, y] = \langle x, My \rangle$$

Then we have $[h, d] = 0$—that is, h is M-conjugate to d. Let P_F be the orthogonal (in the sense of the inner product $[\cdot, \cdot]$) projection onto the linear subspace spanned by d, and let $P_I = I - P_F$. Define the Hilbert space $\mathscr{H}_1 = P_I \mathscr{H}$ with inner product $[\cdot, \cdot]$ and define the operator $M_1 = P_I M$ in $\mathscr{H}_1$.

EXERCISE. Prove that M_1 is a bounded, self-adjoint, positive-definite linear operator from $\mathscr{H}_1$ onto $\mathscr{H}_1$ and that, therefore, h is the unique solution of the equation

$$M_1 x = k_1 \equiv P_1 k$$

Show that the spectral bounds a_1, A_1 of M_1 are related to those a, A of M by $a \leq a_1 \leq A_1 \leq A$. *Hint*: For example, to solve $M_1 x = k'$ for $k' \in \mathscr{H}_1$, let

$$x_\alpha = M^{-1}k' + \alpha M^{-1}d$$

If

$$\alpha = \frac{-\langle k', d \rangle}{\langle d, d \rangle}$$

then $x_\alpha \in \mathscr{H}_1$ and

$$M_1 x_\alpha = P_I M x_\alpha = P_I(k' + \alpha d) = k'$$

If, also, $M_1 x' = k'$ and $x' \in \mathscr{H}_1$, then $P_I M(x_\alpha - x') = 0$, which implies

$$M(x_\alpha - x') = \beta d$$

and

$$0 = [x_\alpha - x', d] = \langle M(x_\alpha - x'), d\rangle = \beta\langle d, d\rangle$$

so $\beta = 0$ and $x_\alpha = x'$.

To solve $M_1 x = k_1$ in $\mathscr{H}_1$, we consider the general form of the CG method obtained by letting

$$K = M_1, \qquad H = M_1^{-2}, \quad \text{so that} \quad N = I, T = M_1$$

All the theory of the CG method applies here, and we can in particular deduce that

$$E_1(x_n) \leq w_n^2 E_1(x_0)$$

where

$$w_n = \frac{2(1-\alpha_1)^n}{(1+\sqrt{\alpha_1})^{2n} + (1-\sqrt{\alpha_1})^{2n}}$$

$$\alpha_1 = \frac{a_1}{A_1}$$

$$E_1(x) = [h - x, h - x] = \langle h - x, M(h - x)\rangle$$

A straightforward calculation shows that the iterates x_i generated by this general algorithm on M_1 in $\mathscr{H}_1$ are precisely the same as the iterates generated by using the standard simple algorithm on M in $\mathscr{H}$ if the initial direction p_0 in the simple algorithm is not chosen as $r_0 = k - Mx_0 = k$ as usual, but by the formula

$$p_0 = P_I r_0 = r_0 + b_{-1} d, \qquad b_{-1} = -\frac{\langle r_0, Md\rangle}{\langle d, Md\rangle}$$

that is, by the usual way of generating CG directions if we identify d with p_{-1}.

EXERCISE. Prove the assertion in the preceding paragraph.

All that the preceding paragraph says is that the standard CG method, modified to require the first direction p_0 to be conjugate to d, is equivalent to a general CG method in a space M-conjugate to d; therefore, the modification of the standard method converges and, in fact, since

$$E_1(x) = [h - x, h - x] = \langle h - x, M(h - x)\rangle \equiv E(x)$$

we have

$$E(x_n) \leq w_n^2 E(x_0)$$

More generally, if we have proceeded through standard CG directions

$p_0, p_1, \ldots, p_{L-1}$ to arrive at $x_L = 0$, then the solution h is M-conjugate to p_i, $0 \leq i \leq L-1$, and we can define P_F as the orthogonal projection (in the $[\cdot, \cdot]$ sense) onto the span of $\{p_0, \ldots, p_{L-1}\}$, $P_I = I - P_F$, $\mathscr{H}_1 = P_I \mathscr{H}$, $M_1 = P_I M$. Then the remainder of the standard CG iterates are precisely the same as those generated by the more general CG method applied to M_1 in $\mathscr{H}_1$ and, therefore, our convergence estimates can make use of the spectral bounds of M_1 on $\mathscr{H}_1$ rather than of M on $\mathscr{H}$. Since the projections P_I are "contracting" as we do this analysis after each new standard CG step, the spectral bounds on the operators M_1 might be contracting, allowing a proof of superlinear convergence. While we have not been successful in accomplishing this, it seems a worthwhile approach.

5.6. CONJUGATE GRADIENTS FOR GENERAL FUNCTIONALS

We now wish to consider minimizing a general functional $f(x)$ over a Hilbert space $\mathscr{H}$ by some analogue of the conjugate-gradient method. In this case, $\nabla f(x)$ plays the role of $2(Mx - k)$ and f''_x plays the role of $2M$. For notational convenience we shall write $J(x) \equiv \nabla f(x)$, $J'_x \equiv f''_x$; we shall also write $r_n \equiv -J(x_n)$, $r \equiv -J(x)$, $J'_n \equiv J'_{x_n}$. Thus, in analogy to the quadratic problem, given x_0, let $p_0 = r_0 = -J(x_0)$; for $n = 0, 1, \ldots$, let $x_{n+1} = x_n + c_n p_n$, c_n to be determined; set $r_{n+1} = -J(x_{n+1})$, and $p_{n+1} = r_{n+1} + b_n p_n$, where

$$b_n = \frac{-\langle r_{n+1}, J'_{n+1} p_n \rangle}{\langle p_n, J'_{n+1} p_n \rangle}$$

If the sequence of vectors p_n that we generate in this manner is admissible, then all the results of Chapter 4 apply to determine the choice of c_n; we consider the admissibility. If we desire

$$\langle r_n, p_n \rangle \geq \alpha \|r_n\|^2, \qquad \alpha > 0$$

then precisely what we need is

$$b_{n-1} \langle r_n, p_{n-1} \rangle \geq -(1 - \alpha) \|r_n\|^2$$

This follows, for example, if

$$|b_{n-1}| \leq (1 - \alpha) \frac{\|r_n\|}{\|p_{n-1}\|}$$

for which

$$\|p_n\| \leq (2 - \alpha) \|r_n\| \quad \text{and} \quad \frac{\langle r_n, p_n \rangle}{\|p_n\|} \geq \frac{\alpha}{2 - \alpha} \|r_n\|$$

However, unless the b_n as determined by the algorithm satisfy such a condition, we must modify b_n and thus lose the relationship with conjugate gradients. Although the study of such methods may be of interest, the rapid convergence of the conjugate-gradient method for quadratic functionals is so desirable in general that we shall limit ourselves to the situation in which similar results can be proved for general functionals. Therefore, we shall now always assume that there exist positive numbers a, A such that

$$aI \leq J'_x \leq AI$$

Thus f is bounded below and tends to infinity with $||x||$. We shall now also determine c_n so as to minimize $f(x_n + cp_n)$ over $c \geq 0$—that is, to solve

$$\langle J(x_n + c_n p_n), p_n \rangle = 0$$

We call this the *pure CG algorithm.* From these conditions it is simple to prove the following [Daniel (1965, 1967b)].

PROPOSITION 5.6.1.

$$\langle p_n, J'_n p_{n-1} \rangle = \langle r_{n+1}, p_n \rangle = 0$$
$$\langle r_n, p_n \rangle = ||r_n||^2$$
$$\langle p_n, J'_n r_n \rangle = \langle p_n, J'_n p_n \rangle$$
$$||p_n||^2 = ||r_n||^2 + b_{n-1}^2 ||p_{n-1}||^2$$
$$\langle r_n, J'_n r_n \rangle = \langle p_n, J'_n p_n \rangle + b_{n-1}^2 \langle p_{n-1}, J'_n p_{n-1} \rangle$$
$$||r_n||^2 \leq ||p_n||^2 \leq \frac{A}{a} ||r_n||^2$$

The following theorem follows from several earlier theorems in Chapters 1 and 4; for clarity we prove it directly here.

THEOREM 5.6.1. The sequence x_n generated by the pure CG algorithm starting with an arbitrary x_0 converges to the unique x^* minimizing f over $\mathscr{H}$. The error estimate

$$||x_n - x^*|| \leq \frac{1}{a} ||J(x_n)||$$

is valid.

Proof: Let $f_n(c) = f(x_n + cp_n)$; then

$$f'_n(c) = \langle J(x_n + cp_n), p_n \rangle$$
$$a\,||p_n||^2 \leq f''_n(c) = \langle J'_{x_n + cp_n} p_n, p_n \rangle \leq A\,||p_n||^2$$

Since

$$f'(0) = -\langle r_n, p_n \rangle = -\|r_n\|^2 < 0$$

we deduce that c_n exists and satisfies

$$c_n \geq \frac{1}{A} \frac{\|r_n\|^2}{\|p_n\|^2} \geq \frac{a}{A^2}$$

Thus, for all $c \leq c_n$, we have for some $0 \leq t \leq 1$,

$$\begin{aligned} f(x_n + cp_n) &= f(x_n) + c\langle J(x_n), p_n \rangle + \frac{c^2}{2} \langle J'_{x_n+tcp_n} p_n, p_n \rangle \\ &\leq f(x_n) - c\|r_n\|^2 + \frac{1}{2} c^2 \frac{A^2}{a} \|r_n\|^2 \end{aligned}$$

Thus

$$\begin{aligned} f(x_{n+1}) &\leq f\left(x_n + \frac{a}{A^2} p_n\right) \\ &\leq f(x_n) - \frac{a}{A^2}\|r_n\|^2 + \frac{1}{2}\frac{a}{A^2}\|r_n\|^2 \leq f(x_n) - \frac{1}{2}\frac{a}{A^2}\|r_n\|^2 \end{aligned}$$

Since $f(x)$ is bounded below, it follows that $J(x_n)$ converges to zero. Since

$$f(x) \geq f(x_0) - \|x - x_0\| \|J(x_0)\| + \tfrac{1}{2}a \|x - x_0\|^2$$

the set of x with $f(x) \leq f(x_0)$ is bounded, hence $\|x_{n+k} - x_n\|$ is bounded; but

$$a\|x_{n+k} - x_n\|^2 \leq \langle J(x_{n+k}) - J(x_n), x_{n+k} - x_n \rangle$$

which converges to zero. Thus there exists x' such that x_n converges to x'; clearly $J(x') = 0$ and $f(x') = \min\{f(x);\, x \text{ in } \mathscr{H}\}$. Uniqueness follows from

$$\|J(x) - J(y)\| \|x - y\| \geq \langle J(x) - J(y), x - y \rangle \geq a\|x - y\|^2$$

as does the error estimate with $x = x'$, $y = x_n$. Q.E.D.

This theorem by itself does not indicate any special value for the method; all of the methods of Chapter 4 behave essentially in this fashion. The advantage of the method for quadratic functions is its rapid convergence rate; we show that, asymptotically, this same rate is obtained in general.

5.7. LOCAL-CONVERGENCE RATES

In examining the local-convergence rate, we discover that estimates can be found simultaneously for a larger class of methods—namely, without choosing b_n via the conjugacy requirement. We assume instead that $\|b_{n-1}p_{n-1}\| \leq D\|r_n\|$ for some D; then

$$\|p_n\|^2 = \|r_n\|^2 + \|b_{n-1}p_{n-1}\|^2 \leq (1 + D^2)\|r_n\|^2$$

which yields

$$\frac{\langle r_n, p_n\rangle}{\|r_n\|\,\|p_n\|} = \frac{\|r_n\|}{\|p_n\|} \geq \frac{1}{(1 + D^2)^{1/2}}$$

so that the p_n are admissible directions. (This assumption can be weakened via Remark 1 following Theorem 4.2.4.) If we examine the effect of this change on the *Proof* of Theorem 5.6.1, we find instead that

$$c_n \geq \frac{1}{A(1 + D^2)}$$

$$f(x_{n+1}) \leq f(x_n) - \frac{1}{2A(1 + D^2)}\|r_n\|^2$$

so that the conclusions of the theorem follow. Thus we have proved the following.

THEOREM 5.7.1. Let $0 < aI \leq J'_x \leq AI$ for $x \in \mathscr{H}$, $J = \nabla f$. Given x_0, let $p_0 = r_0 = -J(x_0)$. For $n = 0, 1, \ldots$, let

$$x_{n+1} = x_n + c_n p_n$$

where c_n solves

$$\langle J(x_n + cp_n), p_n\rangle = 0 \tag{5.7.1}$$

Set

$$r_{n+1} = -J(x_{n+1}), \qquad p_{n+1} = r_{n+1} + b_n p_n$$

where

$$\|b_n p_n\| \leq D\|r_{n+1}\|$$

Then x_n converges to the unique x^* minimizing f over $\mathscr{H}$, and

$$\|x_n - x^*\| \leq \frac{1}{a}\|J(x_n)\|$$

EXERCISE. Supply the details in the *Proof* of Theorem 5.7.1.

EXERCISE. Suppose we only know that $\nabla f(x)$ is Lipschitz-continuous in $\mathscr{H}$ with a fixed Lipschitz constant, and that the algorithm of Theorem 5.7.1 is well defined; prove that $\|\nabla f(x_n)\| \longrightarrow 0$.

We shall analyze the local-convergence properties of this method; we merely note that when

$$b_n = \frac{-\langle r_{n+1}, J'_{n+1} p_n \rangle}{\langle p_n, J'_{n+1} p_n \rangle}$$

we have

$$D = \left(\frac{A}{a} - 1\right)^{1/2}$$

EXERCISE. Prove that

$$D = \left(\frac{A}{a} - 1\right)^{1/2}$$

for the choice of b_n immediately above.

Our approach will be to analyze the convergence in terms of an error measure $E_n(x)$ similar to $E(x)$ in the quadratic case; the work lies in proving that, asymptotically, the convergence is the same for the more general case.

LEMMA 5.7.1.

$$\frac{1}{A(1 + D^2)} \leq \frac{\|r_n\|^2}{A\|p_n\|^2} \leq c_n \leq \frac{1}{a} \frac{\|r_n\|}{\|p_n\|} \leq \frac{1}{a}$$

$$\|c_n p_n\| \leq \frac{\|r_n\|}{a}$$

$$\|r_{n+1}\| \leq \left(1 + \frac{A}{a}\right)\|r_n\|$$

Proof: The lower and upper bounds on c_n follow easily by considering $f_n(c)$ as in the *Proof* of Theorem 5.6.1; since

$$\|r_n\|^2 = \langle r_n, p_n \rangle \leq \|r_n\| \|p_n\|$$

we have

$$\frac{1}{a} \frac{\|r_n\|}{\|p_n\|} \leq \frac{1}{a} \quad \text{and} \quad \|c_n p_n\| \leq \frac{\|r_n\|}{a}$$

Finally,

$$\|r_{n+1}\| \leq \|r_{n+1} - r_n\| + \|r_n\| \leq A\|x_{n+1} - x_n\| + \|r_n\|$$
$$\leq \left(\frac{A}{a} + 1\right)\|r_n\|$$

Q.E.D.

In the quadratic case we found an error measure $E(x)$ such that $E(x_{n+1}) \leq qE(x_n)$ with $q < 1$. We attempt the same here.

DEFINITION 5.7.1.

$$E_n(x) \equiv \langle r, J_n'^{-1} r\rangle, \qquad r \equiv J(x)$$

Note that

$$E_n(x_n) = \langle h_n - x_n, J_n'(h_n - x_n)\rangle$$

where $h_n \equiv x_n + J_n'^{-1} r_n$ is the approximate solution given by Newton's method; thus $E_n(x)$ measures, in a sense, our deviation from that method. We also remark that $E_n(x_n)$ and r_n are of the same order of magnitude—that is,

$$\frac{\|r_n\|^2}{A} \leq E_n(x_n) \leq \frac{\|r_n\|^2}{a}$$

We shall, for convenience, write

$$E_n(x_n) \equiv e_n^2$$

We now assume that there is a constant B such that

$$\|J_x' - J_y'\| \leq B\|x - y\|$$

This assumption only needs to be valid in some neighborhood of the solution x^* since eventually all iterates x_n will be inside that neighborhood.

LEMMA 5.7.2.

$$\frac{\|r_n\|^2}{\langle J_n' p_n, p_n\rangle}\left(\frac{1}{1 + \eta_n}\right) \leq c_n \leq \frac{\|r_n\|^2}{\langle J_n' p_n, p_n\rangle}\left(\frac{1}{1 - \eta_n}\right)$$

where

$$\eta_n = e_n\left[\frac{B\sqrt{A}}{2a^2}\right]$$

Proof: Define

$$g_n(c) = \langle J(x_n + cp_n), p_n \rangle$$
$$= - \| r_n \|^2 + c\langle J'_n p_n, p_n \rangle + c \int_0^1 \langle (J'_{x_n+tcp_n} - J'_n)p_n, p_n \rangle dt$$

This gives

$$-\| r_n \|^2 + c\langle J'_n p_n, p_n \rangle - \tfrac{1}{2}c^2 B \| p_n \|^3 \leq g_n(c) \leq -\| r_n \|^2 + c\langle J'_n p_n, p_n \rangle + \tfrac{1}{2}c^2 B \| p_n \|^3$$

On the interval

$$0 \leq c \leq \frac{1}{a} \frac{\| r_n \|}{\| p_n \|}$$

this implies

$$-\| r_n \|^2 + c\langle J'_n p_n, p_n \rangle - \tfrac{1}{2}c \frac{1}{a} \frac{\| r_n \|}{\| p_n \|} B \| p_n \|^3 \leq g_n(c)$$

and similarly above. Using $aI \leq J'_x$ and $\| r_n \| \leq e_n \sqrt{A}$, we deduce

$$-\| r_n \|^2 + c(1 - \eta_n) \langle J'_n p_n, p_n \rangle \leq g_n(c)$$

and hence derive the upper bound on c_n. The lower bound is derived similarly. Q.E.D.

With Lemma 5.7.2 as a tool, we demonstrate that $E_n(x_n)$ is strongly decreasing, just as was $E(x_n)$ for quadratics.

LEMMA 5.7.3.

$$E_{n+1}(x_{n+1}) - E_n(x_n) \leq -c_n \| r_n \|^2 + de_n^3$$

where

$$d = \frac{B}{a^3}\left(3 + \frac{A}{2a}\right)$$

Proof:

$$E_{n+1}(x_{n+1}) - E_n(x_n) = \langle r_n, (J'^{-1}_{n+1} - J'^{-1}_n) r_n \rangle + \langle r_{n+1} - r_n, J'^{-1}_{n+1} r_{n+1} \rangle + \langle r_{n+1} - r_n, J'^{-1}_{n+1} r_n \rangle \equiv X + Y + Z$$

For the first term,

$$J'^{-1}_{n+1} - J'^{-1}_{n} = J'^{-1}_{n+1}(J'_n - J'_{n+1})J'^{-1}_{n}$$

So we have

$$\| J'^{-1}_{n+1} - J'^{-1}_{n} \| \leq \frac{B}{a^2} \| c_n p_n \| \leq \frac{B}{a^3} \| r_n \|$$

yielding

$$| X | \leq \frac{B}{a^3} \| r_n \|^3$$

For the second term,

$$\begin{aligned} Y &= \langle J(x_n) - J(x_{n+1}), J'^{-1}_{n+1} r_{n+1} \rangle \\ &= \langle J'_{n+1}(x_n - x_{n+1}), J'^{-1}_{n+1} r_{n+1} \rangle + d_1 = d_1 \end{aligned}$$

where, using an integral to represent d_1 as in Lemma 5.7.2, we have

$$| d_1 | \leq \frac{1}{2} \frac{B}{a^3} \| r_n \|^2 \| r_{n+1} \|$$

Using the same device we derive

$$Z = -c_n \| r_n \|^2 + d_2$$

where

$$| d_2 | \leq \frac{1}{2} \frac{B}{a^3} \| r_n \|^3$$

The proof then follows from $\| r_n \| \leq e_n \sqrt{A}$. Q.E.D.

LEMMA 5.7.4.

$$E_{n+1}(x_{n+1}) \leq E_n(x_n)[q + s_n]$$

where

$$q = 1 - \frac{a}{A(1 + D^2)} < 1$$

and $s_n = O(e_n)$ converges to zero. If we use the pure CG method,

$$q = \left(\frac{A - a}{A + a}\right)^2$$

Proof:

$$\begin{aligned} c_n \| r_n \|^2 &\geq \frac{\| r_n \|^2}{\langle J'_n p_n, p_n \rangle} \cdot \frac{\| r_n \|^2}{1 + \eta_n} \\ &= \frac{\| p_n \|^2}{\langle J'_n p_n, p_n \rangle} \cdot \frac{\| r_n \|^2}{\| p_n \|^2} \cdot \frac{\| r_n \|^2}{\langle r_n, J'^{-1}_n r_n \rangle} \cdot \frac{E_n(x_n)}{1 + \eta_n} \\ &\geq \frac{1}{A} \cdot \frac{1}{1 + D^2} \cdot a \cdot \frac{E_n(x_n)}{1 + \eta_n} \end{aligned}$$

Therefore, by the previous lemma,

$$E_{n+1}(x_{n+1}) - E_n(x_n) \leq \frac{-a}{A(1 + D^2)(1 + \eta_n)} E_n(x_n) + d e_n^3$$

so that

$$\begin{aligned} E_{n+1}(x_{n+1}) &\leq E_n(x_n) \left[1 - \frac{a}{A(1 + D^2)(1 + \eta_n)} + d e_n \right] \\ &= E_n(x_n)[q + s_n], \qquad s_n = O(e_n) \end{aligned}$$

For the pure CG method,

$$\langle J'_n p_n, p_n \rangle = \langle J'_n r_n, r_n \rangle + b_{n-1}^2 \langle J'_n p_{n-1}, p_{n-1} \rangle$$

so that

$$\begin{aligned} c_n \| r_n \|^2 &\geq \frac{\| r_n \|^2}{\langle J'_n p_n, p_n \rangle} \cdot \frac{1}{1 + \eta_n} \cdot \frac{E_n(x_n)}{\langle r_n, J'^{-1}_n r_n \rangle} \cdot \| r_n \|^2 \\ &\geq \frac{\langle r_n, r_n \rangle^2}{\langle J'_n r_n, r_n \rangle \langle J'^{-1}_n r_n, r_n \rangle} \cdot \frac{E_n(x_n)}{1 + \eta_n} \\ &\geq \frac{4aA}{(A + a)^2} \cdot \frac{E_n(x_n)}{1 + \eta_n} \end{aligned}$$

by the inequality of Kantorovich [Faddeev–Faddeeva (1963), Kantorovich (1948)]. The remainder follows easily. Q.E.D.

Since

$$E_n(x_n) \geq \frac{\| r_n \|^2}{A} \geq \frac{a^2}{A} \| x_n - x^* \|^2$$

the above lemma completes the *Proof* of the following theorem.

THEOREM 5.7.2. The sequence $\{x_n\}$ generated via Equation 5.7.1 in Theorem 5.7.1 is such that $\| x_n - x^* \|^2$ converges to zero faster than any geometric sequence with convergence factor greater than

$$q = 1 - \frac{a}{A(1 + D^2)}$$

If we use the pure CG method, then

$$q = \left(\frac{A-a}{A+a}\right)^2$$

The above theorem, however, is not a really sharp theorem for the pure CG method, since it does not contain the convergence-rate factor

$$w_n^2 \leq 4\left(\frac{\sqrt{A}-\sqrt{a}}{\sqrt{A}+\sqrt{a}}\right)^{2n}$$

found in the quadratic case. Since the factor

$$\left(\frac{A-a}{A+a}\right)^2$$

is also valid for steepest descent by the same argument made in the *Proof* of Lemma 5.7.4 using the Kantorovich inequality, our CG estimate is no better. We now show that the rate factor w_n^2 is essentially valid here, showing the greater convergence rate for the pure CG method.

THEOREM 5.7.3. For the pure CG algorithm, the following error estimate holds. For any $m \geq 0$ there exists an N_m such that for $n \geq N_m$, we have

$$E_{n+m}(x_{n+m}) \leq (w_m^2 + s_n)E_n(x_n)$$

where $s_n = O(e_n^{1/(4m-3)})$ tends to zero. Here

$$w_m \equiv \frac{2[1-(a/A)]^m}{(1+\sqrt{a/A})^{2m}+(1-\sqrt{a/A})^{2m}} \leq 2\left(\frac{\sqrt{A}-\sqrt{a}}{\sqrt{A}+\sqrt{a}}\right)^m$$

Proof: Consider the iterate x_n and the linear equation $J_n'z = J_n'x_n + r_n$ for z, having solution $h_n \equiv x_n + J_n'^{-1}r_n$. We note that $h_n - x_n$ is J_n'-conjugate to p_{n-1}. If we consider the standard CG method to compute $z = h_n$ starting with $z_0 = x_n$ but requiring that the first direction $\tilde{p}_0$ be J_n'-conjugate to the given direction $d = p_{n-1}$, we have precisely the situation discussed in Section 5.5. Therefore, the sequence of such iterates z_i converges to h_n and

$$\langle h_n - z_m, J_n'(h_n - z_m)\rangle \leq w_m^2\langle r_n, J_n'^{-1}r_n\rangle \tag{5.7.2}$$

The first direction $\tilde{p}_0$ in the modified method is the projection of

$$J_n'x_n + r_n - J_n'z_0 = r_n$$

onto the J_n'-conjugate complement of p_{n-1}; that is,

$$\tilde{p}_0 = p_n$$

If we show that

$$|\langle h_n - z_m, J_n'(h_n - z_m)\rangle - E_{n+m}(x_{n+m})|$$

which equals

$$|\langle h_n - z_m, J_n'(h_n - z_m)\rangle - \langle h_{n+m} - x_{n+m}, J'_{n+m}(h_{n+m} - x_{n+m})\rangle|$$

is of order $e_n^{2+[1/(4m-3)]}$, then we shall have

$$\begin{aligned} E_{n+m}(x_{n+m}) &= \langle h_n - z_m, J_n'(h_n - z_m)\rangle \\ &\quad + [E_{n+m}(x_{n+m}) - \langle h_n - z_m, J_n'(h_n - z_m)\rangle] \\ &\leq [w_m^2 + O(e_n^{1/(4m-3)})]E_n(x_n) \end{aligned}$$

We indicate the proof of the order of magnitude. The sum to be estimated splits into

$$|\langle h_n - z_m, (J_n' - J'_{n+m})(h_n - z_m)\rangle|$$

and

$$|\langle h_n - h_{n+m} + x_{n+m} - z_m, J'_{n+m}(h_n - z_m + h_{n+m} - x_{n+m})\rangle|$$

the first of which is less than

$$B\,||h_n - z_m||^2\,||x_{n+m} - x_n|| = O(e_n^3)$$

by Equation 5.7.2 and the fact that

$$||x_n - x^*|| \leq \frac{\sqrt{A}}{a} e_n$$

Clearly the second part of the sum is less than

$$||h_n - h_{n+m} + x_{n+m} - z_m||\,O(e_n)$$

We estimate the normed term. First,

$$||h_n - h_{n+m}|| = ||\sum_{i=0}^{m-1} (h_{n+i} - h_{n+i+1})||$$

while

$$\begin{aligned} ||h_{j+1} - h_j|| &= ||x_{j+1} - x_j + J'_{j+1}{}^{-1} r_{j+1} - J_j'^{-1} r_j|| \\ &= ||c_j p_j + J'_{j+1}{}^{-1}(r_{n+1} - r_j) + (J'_{j+1}{}^{-1} - J_j'^{-1}) r_j|| = O(e_n^2) \end{aligned}$$

since

$$r_{j+1} - r_j = -J'_{j+1}(c_j p_j) + O(e_n^2)$$

We still must estimate

$$\|x_{n+m} - z_m\| = \|x_{n+m-1} + c_{n+m-1}p_{n+m-1} - z_{m-1} - \tilde{c}_{m-1}\tilde{p}_{m-1}\|$$

where the $\sim$ indicates the z_i-iteration. Since $\tilde{p}_0 = p_n$, an inductive argument [Daniel (1969)] yields

$$\|x_{n+i} - z_i\| = O(e_n^{2-[(4i-4)/(4m-3)]}) \quad \text{or} \quad \|r_{n+i}\| = O(e_n^{1+[1/(4m-3)]})$$

for all i, which leads to

$$\|x_{n+m} - z_m\| = O(e_n^{1+[1/(4m-3)]})$$

Q.E.D.

Thus we have proved that, asymptotically, the rapid convergence of the CG iterates for quadratic functionals carries over to more general functionals. It is in part this convergence, more rapid than any other gradient type of method, that has led to the great popularity of conjugate-gradient methods recently. Of course, so far as the analysis above has been taken, it appears that one must *precisely* compute c_n and make p_n and p_{n-1} *precisely* J'_n-conjugate in order to guarantee convergence. Since such precision is impossible computationally, it is important to know that the rapid-convergence behavior will be maintained under computationally convenient modifications. Much the same results apply, of course, to nearly any method; we consider the methods for which $\|b_{n-1}p_{n-1}\| \leq D\|r_n\|$.

5.8. COMPUTATIONAL MODIFICATIONS

Consider the class of methods given by Equation 5.7.1. The condition that $\langle r_{n+1}, p_n\rangle = 0$—that is, that r_{n+1} be precisely orthogonal to p_n—is very restrictive. Let us consider the algorithm with the sole modification that c_n be chosen so

$$|f_n(c_n)| \equiv \left| \frac{\langle r_{n+1}, p_n\rangle}{\|r_{n+1}\|\,\|p_n\|} \right| \leq \delta$$

for some small $\delta > 0$. Since

$$\begin{aligned} \|r_n\|^2 &= \langle r_n, p_n\rangle + b_{n-1}\langle r_n, p_{n-1}\rangle \\ &\leq \|r_n\|\,\|p_n\| + \delta D\|r_n\|^2 \end{aligned}$$

we have

$$(1 - \delta D)\,||r_n|| \leq ||p_n||$$

LEMMA 5.8.1. If

$$\delta < \frac{1}{1 + 2D}$$

then c_n is bounded away from zero.

Proof:

$$(1 - \delta D)\,||r_n||^2 \leq \langle r_n, p_n\rangle = \frac{\langle r_{n+1}, p_n\rangle}{||r_{n+1}||\,||p_n||}\,||r_{n+1}||\,||p_n|| + \langle r_n - r_{n+1}, p_n\rangle$$
$$\leq \delta\,||p_n||\,||r_{n+1}|| + Ac_n\,||p_n||^2$$

Now

$$||p_n|| \leq (1 + D)\,||r_n|| \quad \text{and} \quad ||r_{n+1}|| \leq ||r_n|| + c_n A\,||p_n||$$

and hence

$$(1 - \delta D)\,||r_n||^2 \leq \delta(1 + D)\,||r_n||\,[||r_n|| + c_n A(1 + D)\,||r_n||]$$
$$+ Ac_n(1 + D)^2\,||r_n||^2$$

which implies

$$c_n \geq \frac{1 - \delta(1 + 2D)}{A(1 + D)^2(1 + \delta)} > 0 \quad \text{if} \quad \delta < \frac{1}{1 + 2D}$$

Q.E.D.

THEOREM 5.8.1. For arbitrary x_0, with $\delta > 0$ small enough (independent of x_0) and c_n and b_n determined as described above, it follows that $x_n \longrightarrow x^*$.

Proof:

$$f(x_n) - f(x_{n+1}) = \langle J(x_{n+1}), -c_n p_n\rangle + \frac{1}{2}\,c_n^2\langle p_n, J'_{x_n + tc_n p_n} p_n\rangle$$
$$\geq c_n\,||p_n||^2\left\{\frac{\langle r_{n+1}, p_n\rangle}{||r_{n+1}||\,||p_n||}\cdot\frac{||r_{n+1}||}{||p_n||} + \frac{1}{2}c_n a\right\}$$
$$\geq c_n^2\,||p_n||^2\left[\frac{1}{2}a - \delta\left(A + \frac{1}{c_n(1 + \delta L)}\right)\right]$$

Because of the lower bound for c_n, if δ is small enough, then

$$f(x_n) - f(x_{n+1}) \geq d_1\,||p_n||^2$$

for some $d_1 > 0$, and hence $\|p_n\| \geq (1 - \delta D)\|r_n\|$ tends to zero, implying $x_n \longrightarrow x^*$. Q.E.D.

The above theorem is somewhat similar to Theorem 4.5.1. In order to obtain good estimates of the local-convergence rate, we need to determine c_n more accurately. According to Lemma 5.7.2, c_n is approximately given by

$$\frac{\|r_n\|^2}{\langle p_n, J'_n p_n\rangle} = \frac{\langle r_n, p_n\rangle}{\langle p_n, J'_n p_n\rangle}$$

Let us consider using this latter value as an approximation $\bar{c}_n$ to c_n, and let us denote the elements of this method by an overbar ($\bar{\ }$)—that is, $\bar{x}_n$, $\bar{p}_n$, etc.—starting with $\bar{x}_0 = x_0$, given. Proceeding for this iteration just as we did in Section 5.7, we can easily find that

$$\|\bar{p}_n\| \leq (1 + D)\|\bar{r}_n\|$$

that c_n solving $\langle r_{n+1}, \bar{p}_n\rangle = 0$ exists and satisfies

$$|c_n - \bar{c}_n| = O(\bar{e}_n)$$

that

$$\|r_{n+1} - \bar{r}_{n+1}\| = O(\bar{e}_n^2)$$

that

$$E_{n+1}(x_{n+1}) \leq q_n E_{\bar{n}}(\bar{x}_n) \quad \text{with} \quad q_n = \left(1 - \frac{a}{A(1 + D^2)}\right) + O(\bar{e}_n) + O(\bar{e}_{n-1})$$

and that

$$E_{\overline{n+1}}(\bar{x}_{n+1}) \leq q_n E_{\bar{n}}(\bar{x}_n)$$

This in essence proves the following proposition.

PROPOSITION 5.8.1. The asymptotic convergence rate for $\|x_n - x^*\|^2$ for the general algorithm with $\|\bar{b}_{n-1}\bar{p}_{n-1}\| \leq D\|\bar{r}_n\|$ and $\bar{c}_n$ determined by its linearized value

$$\bar{c}_n = \frac{\langle \bar{r}_n, \bar{p}_n\rangle}{\langle \bar{p}_n, J'_{\bar{n}}\bar{p}_n\rangle}$$

is greater than that of any geometric sequence with convergence factor greater than

$$q = 1 - \frac{a}{A(1 + D^2)}$$

If $\bar{b}_n = 0$ (steepest descent) or $\bar{b}_n$ is determined by the requirement of $J'_{\bar{n}}$-conjugacy, then

$$q = \left(\frac{A - a}{A + a}\right)^2$$

If we use the conjugacy requirement to determine $\bar{b}_n$, then, as one would expect, the better convergence rate holds [Daniel (1967a, 1969)].

PROPOSITION 5.8.2. If

$$b_n = \frac{-\langle r_{n+1}, J'_{n+1} p_n \rangle}{\langle p_n, J'_{n+1} p_n \rangle} \quad \text{and} \quad c_n = \frac{\langle r_n, p_n \rangle}{\langle p_n, J'_n p_n \rangle}$$

then the asymptotic convergence rate—that is, for e_0 small enough—is described as follows: for every m there exists N_m such that for $n \geq N_m$, we have

$$E_{n+m}(x_{n+m}) \leq [w_m^2 + O(e_n^{1/(4m-3)}) + O(e_{n-1}^{1/(4m-3)})]E_n(x_n)$$

where w_m is given in Theorem 5.7.3.

When $J(x)$ is linear, we know that

$$b_n = \frac{\|r_{n+1}\|^2}{\|r_n\|^2}$$

Since this formula does not involve J'_n in any way, it is computationally useful and has been used in practice for general problems; a computer program can be found in Fletcher–Reeves (1964). If b_n satisfies $\|b_n p_n\| \leq D\|r_{n+1}\|$, then convergence is guaranteed by previous theorems; such an inequality does not appear to be valid in general, however. It can be guaranteed by setting

$$b_n = \min\left\{\frac{\|r_{n+1}\|^2}{\|r_n\|^2}, \frac{A}{a} \cdot \frac{\|r_{n+1}\|}{\|p_n\|}\right\}$$

Another way to compute a b_n which is just as convenient from the computational viewpoint as that above, but more easily analyzed, is via the formula [Poljak (1969a)]

$$b_n = \frac{\langle r_{n+1}, r_{n+1} - r_n \rangle}{\|r_n\|^2}$$

which is a correct formula for quadratics.

EXERCISE. Prove that the three determinations of b_n, namely $\frac{\|r_{n+1}\|^2}{\|r_n\|^2}$, $\frac{\langle r_{n+1}, r_{n+1} - r_n \rangle}{\|r_n\|^2}$, and $\frac{-\langle r_{n+1}, J'_{n+1} p_n \rangle}{\langle p_n, J'_{n+1} p_n \rangle}$, are equivalent on quadratics.

For the global convergence question, we have

$$\begin{aligned}
\|b_n p_n\| &= \|p_n\| \frac{|\langle r_{n+1}, J'_{x_n+\lambda t_n p_n}(x_{n+1} - x_n)\rangle|}{|\langle r_n, p_n\rangle|} \\
&\leq \|p_n\| \frac{At_n \|p_n\| \|r_{n+1}\|}{|\langle r_n - r_{n+1}, p_n\rangle|} \leq \|p_n\| \frac{At_n \|p_n\| \|r_{n+1}\|}{|\langle J'_{x_n+\mu t_n p_n}(x_{n+1} - x_n), p_n\rangle|} \\
&\leq \|p_n\| \frac{At_n \|p_n\| \|r_{n+1}\|}{at_n \|p_n\|^2} \leq \frac{A}{a} \|r_{n+1}\|
\end{aligned}$$

which then implies that we get global convergence. This choice has been used widely in practical computations with optimal-control problems in the Soviet Union [Poljak (1969b, 1969c), Poljak–Skokov (1967a, 1967b), Poljak–Orlov et al. (1967), Poljak–Ivanov–Pukov (1967)]; essentially the same local-convergence results as above are known in this case also [Poljak (1969a)]. In fact, if one uses either of these computationally convenient values of b_n and even the linearized c_n (or, of course, a c_n such that $\langle r_{n+1}, p_n\rangle = 0$), then the rapid local convergence described in Proposition 5.8.2 is valid here also [Daniel (1967a, 1969)]. Thus reasonable computational modifications will preserve the global-convergence properties, while the same asymptotic behavior of the error will occur if the modification is asymptotically exact, such as in the linearizations. It seems reasonable that a criterion such as

$$\left| \frac{\langle r_{n+1}, p_n\rangle}{\|r_{n+1}\| \|p_n\|} \right| < \delta$$

for small fixed δ, with b_n determined by the conjugacy requirement, should lead to the rapid convergence described by w_n; such results do not appear to be known, however.

We have recently learned via private communication with G. Zoutendijk of a global convergence theorem for $b_n = \|r_{n+1}\|^2/\|r_n\|^2$ with no additional modification of b_n, assuming exact minimization along the line.

THEOREM 5.8.2. Let f be bounded below and ∇f be Lipschitz continuous and bounded on $W(x_0)$, the closed convex hull of $\{x; f(x) \leq f(x_0)\}$, and let the conjugate-gradient method with $b_n = \|r_{n+1}\|^2/\|r_n\|^2$ be applied with exact minimization along the line $x_n + cp_n$. Then there exists at least one subsequence x_{n_i} such that $\nabla f(x_{n_i}) \to 0$; if $W(x_0)$ is bounded and f is convex, then the entire sequence $\{x_n\}$ is a minimizing sequence.

Proof: If no subsequence has the stated property, then there are positive numbers B, ϵ, and N such that $\epsilon \leq \|\nabla f(x_n)\| \leq B$ for all $n \geq N$. Now

$$\frac{p_n}{\|r_n\|^2} = \frac{r_n}{\|r_n\|^2} + \frac{p_{n-1}}{\|r_{n-1}\|^2}$$

and hence

$$\left\|\frac{p_n}{\|r_n\|^2}\right\|^2 = \frac{1}{\|r_n\|^2} + \left\|\frac{p_{n-1}}{\|r_{n-1}\|^2}\right\|^2$$

which in turn yields

$$\left\|\frac{p_n}{\|r_n\|^2}\right\|^2 \leq \frac{n-N}{\epsilon^2} + \frac{\|p_N\|^2}{\|r_N\|^4}$$

for $n \geq N$. If we define

$$\alpha_n \equiv \frac{\langle r_n, p_n \rangle}{\|r_n\| \, \|p_n\|} = \frac{\|r_n\|}{\|p_n\|} \geq \frac{1}{B} \frac{\|r_n\|^2}{\|p_n\|}$$

for $n \geq N$, we see that

$$\alpha_n^2 \geq \frac{1}{\dfrac{n-N}{\epsilon^2} + \dfrac{\|p_N\|^2}{\|r_N\|^4}}$$

and hence

$$\sum_{n=0}^{\infty} \alpha_n^2 = \infty.$$

But then, according to Remark 1 after Theorem 4.2.4, this implies that $\|\nabla f(x_n)\| \to 0$, a contradiction. Therefore $\nabla f(x_{n_i}) \to 0$ for some subsequence. Theorem 4.2.1 applied to this subsequence implies that the subsequence is minimizing, while the inequality $f(x_{n_{i+1}}) \leq f(x_j) \leq f(x_{n_i})$ for $n_{i+1} \geq j \geq n_i$ implies that $\{x_n\}$ is a minimizing sequence. Q.E.D.

6 GRADIENT METHODS IN $\mathbb{R}^l$

6.1. INTRODUCTION

Since $\mathbb{R}^l$ under any norm (all of which are equivalent) is a Banach space, and is in fact a Hilbert space under the usual inner-product, all the results of Chapters 4 and 5 apply here. In fact, of course, more detailed results can be obtained for gradient methods in $\mathbb{R}^l$ because of the especially simple structure of this space; in this chapter we examine some of these results.

First, because of the finite dimensionality of $\mathbb{R}^l$, the weak and norm topologies coincide, and any closed, bounded set is (sequentially) compact and vice versa; thus the existence theory of Chapter 1 is simplified, the precise simplifications being left to the reader.

Second, because of the nature of the topology in $\mathbb{R}^l$, criticizing sequences $\{x_n\}$ for a functional f are generally more valuable since, if $W(x_0)$ is bounded (see Section 4.2), then limit points x' of $\{x_n\}$ exist and must be critical points of f; in the following sections we shall examine the consequences of this more closely.

Finally, the asymptotic convergence rates of particular methods can be studied in more detail in $\mathbb{R}^l$; we describe some of these results.

6.2. CONVERGENCE OF $x_{n+1} - x_n$ TO ZERO

We mentioned in Section 4.2, particularly in Theorem 4.2.3, that the convergence of $x_{n+1} - x_n$ to zero could be of great value; in Section 6.3 we shall examine this in some detail. In the present section we shall examine situations in which one can assert that $x_{n+1} - x_n$ does converge to zero.

We have already seen in Chapter 4—according to Theorems 4.6.1, 4.6.2, and 4.6.3—that $x_{n+1} - x_n$ tends to zero when $\{x_n\}$ is determined by use of simple intervals along the line. For the methods of Section 4.7 involving a range function along the line, we could not in general prove that $\|x_{n+1} - x_n\| \longrightarrow 0$, as indicated by Theorems 4.7.1 and 4.7.2 and their extended versions in Corollaries 4.7.1 and 4.7.2. As shown in Corollary 4.7.3, where $p_n = -\nabla f(x_n)$—or, more generally, whenever $\|\nabla f(x_n)\| \longrightarrow 0$ implies $\|p_n\| \longrightarrow 0$—in many special cases of this general method we can assert that $\|x_{n+1} - x_n\| \longrightarrow 0$. It is not true in general, however, that the algorithms of Sections 4.3, 4.4, and 4.5 involving minimization along the line necessarily yield $\|x_{n+1} - x_n\| \longrightarrow 0$; contradicting examples can be created. We can, however, show that for many methods and certain kinds of functions we must always have $\|x_{n+1} - x_n\| \longrightarrow 0$.

If $W(x_0)$ is compact, then $\{x_n\}$ has limit points x' and $\nabla f(x') = 0$. Hence the following proposition follows.

PROPOSITION 6.2.1. If ∇f is continuous on the compact set $W(x_0)$ and $\nabla f(x) = 0$ has only one solution x^*, then $x_n \longrightarrow x^*$.

We seek more significant results.

THEOREM 6.2.1 [Elkin (1968)]. If $W(x_0)$ is compact, if there exists a $\delta > 0$ such that

$$f(x_{n+1}) \leq f[tx_{n+1} + (1 - t)x_n] \leq f(x_n) \quad \text{for} \quad 0 \leq t \leq \delta \quad \text{for all} \quad n$$

and if f is not constant on any line segments in $W(x_0)$, then $\|x_{n+1} - x_n\| \longrightarrow 0$.

Proof: If $\|x_{n+1} - x_n\| \nrightarrow 0$, then we may assume that $x_{n_i} \longrightarrow x'$, $x_{n_i+1} \longrightarrow x''$, $x' \neq x''$ for some subsequence n_i. Thus

$$f(x_{n_{i+1}}) \leq f(x_{n_i+1}) \leq f[tx_{n_i+1} + (1 - t)x_{n_i}] \leq f(x_{n_i}) \quad \text{for} \quad 0 \leq t \leq \delta$$

which implies

$$f(x') \leq f(x'') \leq f[tx'' + (1 - t)x'] \leq f(x') \quad \text{for} \quad 0 \leq t \leq \delta$$

which means that f is constant on a line segment. Q.E.D.

THEOREM 6.2.2. If x_{n+1} is determined as in Theorem 4.4.1, if ∇f is continuous on $W(x_0)$, and if f is not constant on any line segment in the compact set $W(x_0)$, then $\|x_{n+1} - x_n\| \longrightarrow 0$.

Proof: In the *Proof* of Theorem 4.4.1 we observed that t_n provides the global minimum of

$$f(x_n + tp_n) - \alpha_n t\langle\nabla f(x_n), p_n\rangle \quad \text{for} \quad 0 \leq t \leq t_n$$

Since $\|x_{n+1} - x_n\|$ is bounded because $W(x_0)$ is compact, we know from Theorem 4.4.1 that

$$\langle \nabla f(x_n), x_{n+1} - x_n \rangle = \left\langle \nabla f(x_n), \frac{p_n}{\|p_n\|} \right\rangle \|x_{n+1} - x_n\| \longrightarrow 0$$

If we have $x_{n_i} \longrightarrow x'$, $x_{n_i+1} \longrightarrow x'' \neq x'$, then

$$\begin{aligned} f(x_{n_{i+1}}) - \alpha_{n_i}\langle \nabla f(x_{n_i}), x_{n_i+1} - x_{n_i}\rangle &\leq f(x_{n_i+1}) - \alpha_{n_i}\langle \nabla f(x_{n_i}), x_{n_i+1} - x_{n_i}\rangle \\ &\leq f[\lambda x_{n_i} + (1-\lambda)x_{n_i+1}] \\ &\quad - \alpha_{n_i}(1-\lambda)\langle \nabla f(x_{n_i}), x_{n_i+1} - x_{n_i}\rangle \\ &\leq f(x_{n_i}) \quad \text{for} \quad 0 \leq \lambda \leq 1 \end{aligned}$$

which yields

$$f(x') \leq f(x'') \leq f[\lambda x' + (1-\lambda)x''] \leq f(x')$$

a contradiction to the assumptions about f. Q.E.D.

The above results treat the methods of Section 4.4; we still have not considered the method of Section 4.3, in which x_{n+1} minimizes f along the line $x_n + tp_n$. We only know how to treat this via more general results applying to all methods.

THEOREM 6.2.3. If $W(x_0)$ is compact, if $f(x_{n+1}) \leq f(x_n)$ for all n, if $\|\nabla f(x_n)\| \longrightarrow 0$, and if there exists a function $\delta(t)$ for $t \geq 0$, $\delta(t) \geq 0$, with $\delta(t_n) \longrightarrow 0$ if and only if $t_n \longrightarrow 0$ and satisfying

$$|f(x) - f(y)| + \|\nabla f(x) - \nabla f(y)\| \geq \delta(\|x - y\|), \quad \text{for} \quad x, y \in W(x_0)$$

then $\|x_{n+1} - x_n\| \longrightarrow 0$.

Proof: If $\|x_{n+1} - x_n\| \not\rightarrow 0$, we take $x_{n_i} \longrightarrow x'$, $x_{n_i+1} \longrightarrow x'' \neq x'$. Then clearly, $f(x') = f(x'')$. Thus

$$\delta(\|x_{n_i+1} - x_{n_i}\|) \leq |f(x_{n_i+1}) - f(x_{n_i})| + \|\nabla f(x_{n_i+1}) - \nabla f(x_{n_i})\| \longrightarrow 0$$

Q.E.D.

We recall that if f is strictly convex and ∇f is continuous, then

$$f(x_2) - f(x_1) > \langle x_2 - x_1, \nabla f(x_1)\rangle \quad \text{if} \quad x_1 \neq x_2$$

Thus, if $\langle x_2 - x_1, \nabla f(x_1)\rangle \geq 0$, we conclude $f(x_2) > f(x_1)$; a function satisfying this property is called *strictly pseudo-convex* [Elkin (1968), Ponstein (1967)].

THEOREM 6.2.4 [Elkin (1968)]. For all x, y in the compact set $W(x_0)$, let $\langle x - y, \nabla f(y)\rangle \geq 0$ imply $f(x) > f(y)$. Let $f(x_{n+1}) \leq f(x_n)$ for all n, let ∇f be continuous in $W(x_0)$, and let $\langle \nabla f(x_n), x_{n+1} - x_n\rangle \to 0$. Then $\|x_{n+1} - x_n\| \to 0$.

Proof: As usual, we take $x_{n_i} \to x'$, $x_{n_i+1} \to x'' \neq x'$ if $\|x_{n+1} - x_n\| \nrightarrow 0$. Of course, $f(x') = f(x'')$ and, therefore,

$$\langle x'' - x', \nabla f(x')\rangle < 0$$

by the strict–pseudo-convexity assumption. However, $\langle x_{n_i+1} - x_{n_i}, \nabla f(x_{n_i})\rangle$ converges to $\langle x'' - x', \nabla f(x')\rangle$ by continuity and to zero by assumption, a contradiction. Q.E.D.

Thus we have found a large variety of ways to guarantee that $\|x_{n+1} - x_n\| \to 0$; let us now see how this restricts the nature of the limit set of $\{x_n\}$—that is, the set of limit points.

6.3. THE LIMIT SET OF $\{x_n\}$

Let L denote the (closed) set of limit points of the sequence $\{x_n\}$. We next study the nature of L in terms of the sequence $\{x_n\}$, particularly under the assumption that $\|x_{n+1} - x_n\| \to 0$. First we strengthen Theorem 4.2.3; recall that a *continuum* is a closed set which cannot be written as the union of two nonempty, disjoint, closed sets.

THEOREM 6.3.1 [Ostrowski (1966a, b)]. If the sequence $\{x_n\}$ in $\mathbb{R}^l$ is bounded, if $\|x_{n+1} - x_n\| \to 0$, and if $\{x_n\}$ does not converge, then the limit set L is a continuum.

Proof: Suppose we can write the closed set L as $L = C_1 \cup C_2$ where C_1 and C_2 are closed, nonempty, and $C_1 \cap C_2 = \phi$. Then there is an $\epsilon > 0$ such that $\|c_1 - c_2\| \geq \epsilon$ for all $c_1 \in C_1, c_2 \in C_2$. For $n \geq N_\epsilon$, we have $\|x_{n+1} - x_n\| \leq \epsilon$. Choose c_1 in C_1; there exist arbitrarily large $n \geq N_\epsilon$ with

$$\|x_n - c_1\| \geq \frac{\epsilon}{3}$$

For such n there exist $m > n$ with

$$\|x_m - c_{2,m}\| \leq \frac{2\epsilon}{3}$$

for some $c_{2,m}$ in C_2. Let m_0 be the smallest such index. Then

$$\|x_{m_0-1} - c_2\| > \frac{2\epsilon}{3}$$

for all c_2 in C_2 and hence, since

$$\| x_{m_0} - x_{m_0-1} \| \leq \frac{\epsilon}{3}$$

we have

$$\| x_{m_0} - c_2 \| > \frac{\epsilon}{3}$$

for all c_2 in C_2. Thus we have

$$\frac{\epsilon}{3} < \| x_{m_0} - c_2 \| \quad \text{for all} \quad c_2 \in C_2 \quad \text{and} \quad \| x_{m_0} - c_{2,m_0} \| \leq \frac{2\epsilon}{3}$$

If we do this for an infinite sequence of indices m_0, there is a limit point x' of the sequence, so $x' \in L$ and the distance of x' from C_2 is at least $\epsilon/3$ since

$$\frac{\epsilon}{3} < \| x_{m_0} - c_2 \|$$

for all $c_2 \in C_2$ so x' must lie in C_1. Yet its distance from C_2 is at most $2\epsilon/3$ since

$$\| x_{m_0} - c_{2,m_0} \| \leq \frac{2\epsilon}{3}$$

while C_1 and C_2 are ϵ apart—a contradiction. Q.E.D.

The import of the above result should be quite obvious; if ∇f is continuous, then $\nabla f(x) = 0$ for all x in the limit set L, if $\| \nabla f(x_n) \| \to 0$, and therefore we can conclude that $\{x_n\}$ must be convergent if $\| x_{n+1} - x_n \| \to 0$ and if $\{x; \nabla f(x) = 0\}$ contains no continuum as a subset. Under some additional hypotheses on the method of generating $\{x_n\}$, it is possible to discover still more about the properties of the limit set L, following Ostrowski (1966a); if these properties are not valid on $\{x; \nabla f(x) = 0\}$, then we again conclude that $\{x_n\}$ is convergent. We consider some of these results.

We assume that

$$x_{n+1} = x_n + t_n p_n \tag{6.3.1}$$

where

$$\left.\begin{array}{lll} \delta_n \equiv \| t_n p_n \| \leq R \| \nabla f(x_n) \|, & R < \infty & \\ f(x_n) - f(x_{n+1}) \geq r \| \nabla f(x_n) \|^2, & r > 0, & \| \nabla f(x_n) \| \to 0 \end{array}\right\} \tag{6.3.2}$$

These assumptions are valid, for example, for the simple-interval methods of

Section 4.5, and for other methods under some added hypotheses. It is, for instance, valid for the following form of the method of Section 4.6 using a range function. Let $d(t) = \delta t$, $0 < \delta < \frac{1}{2}$, $p_n = ||\nabla f(x_n)||q_n$, $||q_n|| = 1$,

$$\langle -\nabla f(x_n), p_n \rangle \geq \epsilon ||\nabla f(x_n)||^2, \qquad \epsilon > 0$$

and suppose that

$$||\nabla f(x) - \nabla f(y)|| \leq L||x - y||$$

for x, y in $W(x_0)$. Then, by Corollary 4.6.1, we can apply the method of Theorem 4.6.1 and find that $||\nabla f(x_n)|| \longrightarrow 0$; since $t_n \leq 1$, we also have $||x_{n+1} - x_n|| \longrightarrow 0$ and

$$\delta_n \equiv ||t_n p_n|| \leq R||\nabla f(x_n)||, \qquad R = 1$$

We also have that

$$f(x_n) - f(x_{n+1}) \geq \delta t_n \langle -\nabla f(x_n), p_n \rangle \geq \delta \epsilon t_n ||\nabla f(x_n)||^2$$

Thus, if t_n is bounded away from zero,

$$f(x_n) - f(x_{n+1}) \geq r||\nabla f(x_n)||^2$$

with $r > 0$ as asserted; we show that t_n is bounded from zero. If not, suppose $t_n \longrightarrow 0$ (actually a subsequence). Since $t_n \neq 1$, we know that t_n has been chosen so that

$$\delta \leq |g(x_n, t_n, p_n) - 1|$$

Then

$$\delta \leq \left| \frac{f(x_n) - f(x_{n+1})}{t_n \langle -\nabla f(x_n), p_n \rangle} - 1 \right|$$
$$\leq \left| \frac{\langle -\nabla f(x_n + \lambda_n t_n p_n) + \nabla f(x_n), t_n p_n \rangle}{t_n \langle -\nabla f(x_n), p_n \rangle} \right. \quad \text{for some} \quad \lambda_n \in (0, 1)$$

which yields

$$\delta \langle -\nabla f(x_n), p_n \rangle \leq ||p_n|| \, ||\nabla f(x_n) - \nabla f(x_n + \lambda_n t_n p_n)|| \leq L t_n ||p_n||^2$$

Therefore,

$$t_n \geq \frac{\delta \langle -\nabla f(x_n), p_n \rangle}{L||p_n||^2} \geq \frac{\delta \epsilon ||\nabla f(x_n)||^2}{L||\nabla f(x_n)||^2} = \frac{\delta \epsilon}{L} > 0$$

which is a contradiction. Thus the assumptions of Equations 6.3.1 and 6.3.2 are valid for this important method.

EXERCISE. By the same kind of argument as above, show that Equations 6.3.1 and 6.3.2 are also valid for a functional f and directions p_n as described above if t_n is chosen as the first local minimum or global minimum of f along $x_n + tp_n$.

Thus our assumptions in Equations 6.3.1 and 6.3.2 are valid for a large class of methods; we now consider the implications of these assumptions. Without loss of generality, we assume $f(x_n) \longrightarrow 0$.

LEMMA 6.3.1. Assume that z is a limit point of $\{x_n\}$ and that for $||x - z|| \leq \rho > 0$ we have

$$f(x) \leq \Gamma ||\nabla f(x)||^2, \qquad \Gamma < \infty$$

Then $\{x_n\}$ converges to z.

Proof: Without loss of generality, we take $z = 0$. Define

$$Q = \max\left(\frac{\Gamma}{r}, 1\right)$$

D = the greatest integer less than or equal to $4Q^2$. Suppose we have an integer m and an integer $p > D$ such that $||x_{m+s}|| \leq \rho$ for $s = 0, 1, \ldots, p$. we write

$$r\sum_{i=s}^{p} ||\nabla f(x_{m+i})||^2 \leq f(x_{m+s}) - f(x_{m+p+1}) \leq f(x_{m+s}) \leq \Gamma ||\nabla f(x_{m+s})||^2$$

for s $= 0, 1, \ldots, p$, since $||x_{m+s}|| \leq \rho$. Thus we have

$$\sum_{i=s}^{p} u_i \leq Qu_s, \qquad Q \geq 1, \qquad u_i = ||\nabla f(x_{m+i})||^2, \qquad s = 0, 1, \ldots, p$$

Solving this inequality [Ostrowski (1966a)] yields

$$u_{i+D} < \tfrac{1}{4}u_i, \quad \text{i.e.,} \quad ||\nabla f(x_{m+i+D})|| < \tfrac{1}{2}||\nabla f(x_{m+i})||, \qquad i = 0, 1, \ldots, p - D$$

Now, since the origin z is a limit point of $\{x_n\}$ and $||\nabla f(x_n)|| \longrightarrow 0$ and $||x_{n+1} - x_n|| = \delta_n \longrightarrow 0$, there exists a fixed m depending on ρ such that

$$||x_m|| \leq \frac{\rho}{3}$$

$$||x_{m+s}|| \leq \rho \quad \text{for} \quad s = 1, 2, \ldots, D + 1$$

and

$$\sum_{i=0}^{D-1} ||\nabla f(x_{m+i})|| \leq \frac{\rho}{3R}$$

We now show that $\|x_n\| \leq \rho$ for all $n \geq m$, which, since ρ can be taken arbitrarily small, gives $\|x_n - z\| \longrightarrow 0$. If not, let x_{m+p+1} be the first such x with $\|x_{m+p+1}\| > \rho$; then $\|x_{m+p}\| \leq \rho$ and $p > D$ so that m and p are allowable values for the inequality found above. Thus we have

$$\begin{aligned}\|x_{m+p+1} - x_m\| &= \sum_{i=0}^{p} \|x_{m+i+1} - x_{m+i}\| \leq R \sum_{i=0}^{p} \|\nabla f(x_{m+i})\| \\ &\leq R\left\{\sum_{i=0}^{D-1} \|\nabla f(x_{m+i})\| + \sum_{D}^{2D-1} \|\nabla f(x_{m+i})\| + \dots\right. \\ &\leq R \sum_{i=0}^{D-1} \|\nabla f(x_{m+i})\| \{1 + \tfrac{1}{2} + \tfrac{1}{4} + \dots\} \leq \frac{2\rho}{3}\end{aligned}$$

Thus

$$\rho < \|x_{m+p+1}\| \leq \|x_m\| + \|x_{m+p+1} - x_m\| \leq \frac{\rho}{3} + \frac{2\rho}{3} \leq \rho$$

which is a contradiction. Q.E.D.

Now we can use the above to prove a theorem on the nature of the limit set L.

THEOREM 6.3.2 (Ostrowski (1966a)]. If f is twice continuously differentiable on $W(x_0)$, if z is a limit point of $\{x_n\}$, and if the Jacobian matrix f''_z of f at z is nonsingular, then $\{x_n\}$ converges to z.

Proof: Since $f(z) = 0$ and $\nabla f(z) = 0$, near z we have

$$f(x) = \langle f''_z(x - z), x - z\rangle + o(\|x - z\|)^2$$

Without loss of generality, we let $z = 0$. Since f''_z is symmetric and nonsingular, we order its eigenvalues $0 < |\lambda_1| \leq \cdots \leq |\lambda_l|$ and we can choose a norm such that $\|f''_z\| = |\lambda_l|$. Also,

$$\begin{aligned}\|\nabla f(x)\|^2 &= \|\nabla f(x) - \nabla f(z)\|^2 \\ &= \|f''_z(x - z)\|^2 + o(\|x - z\|^2) \\ &\geq [\lambda_1^2 + o(1)]\,\|x - z\|^2\end{aligned}$$

and thus

$$f(x) \leq \left[\frac{|\lambda_l|}{\lambda_1^2} + o(1)\right] \|x\|^2$$

so that the hypotheses of Lemma 6.3.1 are satisfied. Q.E.D.

The import of the above is that, if f'' is not singular on any continuum, then we conclude that $\{x_n\}$ converges, since otherwise L is a continuum and

f'' must be singular on it. Actually, in Ostrowski (1966a) it is proved that if f is four times continuously differentiable in $W(x_0)$, then, if for some particular z in L the rank of f''_z is $l - 1$, it follows that $\{x_n\}$ converges to z, providing more detail for the theorem above. Thus we see that in $\mathbb{R}^l$, if the assumptions in Equations 6.3.1 and 6.3.2 are valid—as they are for many methods—the sequence $\{x_n\}$ is convergent except for very pathological functionals f whose gradient and second-derivative matrix "vanish" to a high degree on a continuum.

6.4. IMPROVED CONVERGENCE RESULTS

In Section 5.7 we derived a local-convergence-rate estimate for methods in which

$$p_{n+1} = r_{n+1} + b_n p_n, \qquad r_{n+1} = -\nabla f(x_{n+1})$$

and $\|b_n p_n\| \leq D \|r_{n+1}\|$ for some constant D. In particular, for the pure steepest-descent algorithm with $b_n \equiv 0$, we found that the convergence was at least as fast as a geometric sequence with convergence ratio $(A - a)/(A + a)$, where

$$0 < aI \leq f''_x \leq AI$$

For the steepest-descent method in $\mathbb{R}^l$ this is in fact the best possible result in general. To see this, one considers the simple case of a quadratic functional

$$f(x) = \langle x, Mx \rangle$$

with

$$0 < aI \leq M \leq AI$$

We know that if one proceeds along $x_n + tp_n$ to the minimum in that direction, then $\langle r_{n+1}, p_n \rangle = 0$; in $\mathbb{R}^2$ this implies that one uses precisely the same two directions r_0 and r_1 continually. Asymptotically this is true in $\mathbb{R}^l$ for $l \geq 2$ as well. More precisely [Akaike (1959), Forsythe (1968)], asymptotically the directions alternate between two fixed directions in the space spanned by a certain two eigenvectors of M and thus the convergence is *precisely* linear and described by the ratio $(A - a)/(A + a)$ for any iteration not starting with an eigenvector as x_0. Essentially the same results have been found for the s-dimensional optimum-gradient method [Forsythe (1968)], in which x_{n+1} is chosen to minimize f over the s-dimensional plane

$$x_n + \sum_{i=1}^{s} \alpha_i M^i x_n$$

Since no better results can possibly hold for more general functionals, we see that the rate given for the steepest-descent method is the best possible.

A better estimate, however, can be given for the conjugate-gradient method also discussed in Section 5.7; in that case, we know by Theorem 5.2.2 that for quadratic functionals in $\mathbb{R}^l$ the precise solution is found in at most l steps—that is, x_l is the solution. Therefore, we must look at general functionals in $\mathbb{R}^l$ to find a better convergence estimate; the estimate is provided by using Theorem 5.7.3, which states that the convergence rate is essentially that given for quadratics by considering the method as an optimal process. If we use the symbol D to represent arbitrary constants, then in the proof of that theorem we found

$$||x_{n+m} - z_m|| \leq D||r_n||^{1+[1/(4m-3)]}$$

Thus

$$||x_{n+m} - x^*|| \leq ||x_{n+m} - z_m|| + ||z_m - h_n|| + ||h_n - x^*||$$

where $h_n = x_n - J_n'^{-1}J(x_n)$ is the Newton step and z_m is the point obtained in m steps of the conjugate-gradient method used to compute h_n by solving $J'_n z = J'_n x_n - J(x_n)$. If we are in $\mathbb{R}^l$ and $m = l$, then we know that $z_l = z_m = h_n$, and thus

$$\begin{aligned}
||x_{n+l} - x^*|| &\leq ||x_{n+l} - z_l|| + ||h_n - x^*|| \\
&\leq D||J(x_n)||^{1+[1/(4l-3)]} + ||h_n - x^*|| \\
&\leq D||J(x_n) - J(x^*)||^{1+[1/(4l-3)]} + ||x_n - x^* - J_n'^{-1}[J(x_n) - J(x^*)]|| \\
&\leq D||x_n - x^*||^{1+[1/(4l-3)]}
\end{aligned}$$

Thus we have proved the following theorem.

THEOREM 6.4.1. If $0 < aI \leq J'_x \leq AI$, and $||J'_x - J'_x|| \leq B||x - y||$, then the pure conjugate-gradient (CG) method in $\mathbb{R}^l$ yields a sequence $\{x_n\}$ converging to the point x^* minimizing f and the asymptotic-convergence rate is described by

$$||x_{n+l} - x^*|| \leq D||x_n - x^*||^{1+[1/(4l-3)]}, \qquad D \text{ constant}$$

Since l steps of this method yield the same error-reduction factor as one step of a method with superlinear—actually, $\{1 + [1/(4l - 3)]\}$th order—convergence, we call this $(1/l)$-superlinear convergence. It would seem likely from this result and from our later Theorems 7.4.3 and 7.4.5 that the convergence is in fact at least superlinear—that is, that

$$\lim_{n\to\infty} \frac{||x_{n+1} - x^*||}{||x_n - x^*||} = 0$$

and probably that

$$\lim_{n\to\infty} \frac{\| x_{n+1} - x^* \|}{\| x_n - x^* \|^p} = 0, \qquad p = \left(1 + \frac{1}{4l - 3}\right)^{1/l}$$

Neither of these results has been proved or disproved, however.

At the end of Section 5.8 we considered the possibility of using the conjugate-gradient method with b_n determined by the formula

$$b_n = \frac{\| r_{n+1} \|^2}{\| r_n \|^2}$$

as in the quadratic case; for this method, all of the convergence results, including the $1/l$-superlinear convergence described above, are valid near the solution and we have global convergence. Practical experience in $\mathbb{R}^l$ has shown that one must periodically restart the algorithm with a steepest-descent direction; that is, one should let

$$b_n = 0 \quad \text{if} \quad n = 0 \quad (\text{modulo } m) \quad \text{for some } m$$

Commonly $m = l$ or $m = l + 1$. We analyze this method.

THEOREM 6.4.2 [Ortega–Rheinboldt(1968)]. Let $0 < aI \leq f''_x \leq AI < \infty$ for all x in $\mathbb{R}^l$, and let $\{x_n\}$ be determined by

$$\begin{aligned} x_{n+1} &= x_n + c_n p_n, \qquad f(x_n + c_n p_n) = \min_{c \gtrless 0} f(x_n + c p_n) \\ p_n &= r_n + b_{n-1} p_{n-1}, \\ b_{n-1} &= 0 \quad \text{if} \quad n - 1 = 0 \quad (\text{modulo } m) \quad \text{and} \quad b_{n-1} = \frac{\gamma_n}{\gamma_{n-1}} \end{aligned}$$

otherwise,

$$\gamma_n = \| r_n \|^2$$

Then $\| x_n - x^* \| \longrightarrow 0$ and all of the convergence estimates for the conjugate-gradient method hold near x^*.

Proof: For any $n \geq 0$ let n_0 be the greatest integer not exceeding n which is congruent to zero modulo m. Then it is easy to see that

$$p_n = \sum_{i=n_0}^{n} \frac{\gamma_n}{\gamma_i} r_i$$

Since $W_0 \equiv \{x; f(x) \leq f(x_0)\}$ is compact in $\mathbb{R}^l$, we can define $C < \infty$ with $C = \max_{W_0} \| \nabla f(x) \|$. Also let

$$\gamma_{i_n} = \min_{n_0 \leq i \leq n} \gamma_i$$

Then

$$\|p_n\| \leq \sum_{i=n_0}^{n} \frac{\gamma_n}{\gamma_i} \|r_i\| \leq \frac{(m+1)C}{\gamma_{i_n}} \gamma_n$$

Therefore,

$$\langle r_n, p_n \rangle = \|r_n\|^2 = \gamma_n \geq \frac{\|p_n\|}{C(m+1)} \gamma_{i_n}$$

$$\left\langle r_n, \frac{p_n}{\|p_n\|} \right\rangle \geq \frac{\gamma_{i_n}}{C(m+1)}$$

Proceeding as in Theorem 4.3.1, we conclude that

$$\left\langle r_n, \frac{p_n}{\|p_n\|} \right\rangle \longrightarrow 0$$

and hence

$$\gamma_{i_n} = \|r_{i_n}\|^2 \longrightarrow 0$$

However, since

$$\left\langle r_n, \frac{p_n}{\|p_n\|} \right\rangle \longrightarrow 0$$

and W_0 is bounded, we conclude from Theorem 6.2.4 that $\|x_{n+1} - x_n\| \to 0$. Since ∇f is uniformly continuous in W_0 it follows then that $\|\nabla f(x_n) - \nabla f(x_{i_n})\| \to 0$ and hence

$$\|\nabla f(x_n)\| \longrightarrow 0$$

This then implies $\|x_n - x^*\| \to 0$. Once the algorithm essentially restarts [$n = 0$ (modulo m)] near enough to x^*, all of the original conjugate-gradient theory applies to give the convergence results. Q.E.D.

> **EXERCISE.** Give another, simpler proof of the above theorem by showing, first, that the sequence $\{z_i\}$ converges to x^*, where $z_i \equiv x_{i \cdot m}$ is essentially generated by the steepest-descent method; and then by showing that this implies the convergence of the entire sequence $\{x_n\}$.

Theorem 6.4.2 was first proved in Ortega–Rheinboldt (1968), where it is given, since the local-rate-of-convergence estimates were not desired there, under more general assumptions on f—namely, that W_0 be compact and that some condition guaranteeing $\|x_{n+1} - x_n\| \to 0$ be satisfied. The theorem is valuable from a computational viewpoint, since the conjugate-gradient method is known to exhibit its powerful convergence properties thereby, even when implemented in a fashion not requiring explicit use of the Hessian f'' of second derivatives.

EXAMPLE [Daniel (1967a)]. Consider obtaining a solution to $\nabla^2 u = 16yz(e^{yzu} - 1)$ in $[0, 1] \times [0, 1]$ with $u(y, z) = 0$ on the boundary, having unique solution $u(y, z) \equiv 0$, using the usual five-point formula with an $h \times h$ mesh. The discretized equations are just $\nabla f(x) = 0$ for some uniformly convex functional f, with x in $\mathbb{R}^l$, $l = [(1/h) - 1]^2$. Using (1) the pure CG method with b_n determined for exact conjugacy; (2) the modified CG method with

$$b_n = \min\left\{\frac{\|r_{n+1}\|^2}{\|r_n\|^2}, \frac{A}{a} \cdot \frac{\|r_{n+1}\|}{\|p_n\|}\right\}$$

and (3) steepest descent—that is, $b_n \equiv 0$—the number of iterations and computer time required to reduce $\|\nabla f(x)\|$ from 100 to 10^{-6} on an Electrologica X1, a very slow machine, for $h = \frac{1}{4}$, were respectively (1) 12 iterations, 210 seconds; (2) 13 iterations, 211 seconds; and (3) 40 iterations, 454 seconds.

6.5. CONSTRAINED PROBLEMS

Most of the previous comments of this chapter apply to problems with constraints; the topology of $\mathbb{R}^l$ simplifies convergence questions. We shall not attempt, however, to look into the details of these specializations.

The methods discussed in Section 4.10 are of course applicable in $\mathbb{R}^l$ and, in fact, usually originated there. In particular, the methods of Theorems 4.10.1 and 4.10.2 are extensions of methods in Frank–Wolfe (1956), Gilbert (1966), and Rosen (1960–61). Once we restrict ourselves to $\mathbb{R}^l$ and constrained problems, all of the complex theory of mathematical programming and its many algorithms presents itself. Since we could not hope to proceed to any real depth of presentation of this material in this text, we go no further with mathematical-programming methods but rather refer the reader to the literature [Fiacco–McCormick (1968), Hadley (1964), Mangasarian (1969), Zangwill (1969)].

6.6. MINIMIZATION ALONG THE LINE

To turn a theoretical method into a useful computational algorithm, one needs to be able to implement all steps of the method reasonably quickly and accurately. The simple-interval method of Section 4.5 in theory requires a knowledge of the Lipschitz constant L; in practice, of course, one would usually try some such procedure as letting t_n equal the first of the numbers $T, \alpha T, \alpha^2 T, \ldots, \alpha \in (0, 1)$, for which f decreases significantly. We saw how this approach could be justified in Section 4.6, where it was applied to the method there for finding t_n. We have not, however, indicated how one might implement the methods of Sections 4.3 and 4.4, except for the material concerning the search routine in Section 4.7. It is very difficult to say how one should proceed with the approximate minimization along the line

$x_n + tp_n$. Clearly one need not waste time doing this too accurately "far away" from the solution, but one does demand accuracy "near" the solution. These are difficult terms to define, but one might reasonably use $\|\nabla f(x_n)\|$ as a measure, if all the variables in f are scaled so as to have essentially the same importance; such scaling is always important computationally. We shall assume that such questions of needed accuracy can be answered and shall proceed with a presentation of methods for acquiring this accuracy; we shall present methods which appear from practice to give satisfactory accuracy at a reasonable cost of efficiency.

Most algorithms in use for finding a minimum along a line rely on an iterative interpolation method rather than direct search; they do, however, often incorporate as a first step a preliminary search to isolate the minimizing point in a certain interval. Therefore, we shall first look briefly at the results of direct-search methods. We have already considered in Section 4.7 how one can search to locate the minimizing point for a strongly quasi-convex functional.

Although we generally prefer interpolation methods for accurate determination of the minimizing point in practice, we describe a direct-search method for finding the minimizing point as accurately as possible. Suppose that the minimizing point for a strongly quasi-convex function $g(t) = f(x_n + tp_n)$ of a real variable t is known to lie in the interval $[a_0, b_0]$. If we insert two points, $a_0 < t_{0,1} < t_{0,2} < b_0$, and evaluate g there, then the minimizing point is in $[a_0, t_{0,2}]$ if $g(t_{0,1}) < g(t_{0,2})$; in $[t_{0,1}, b_0]$ if $g(t_{0,1}) > g(t_{0,2})$; and in $[t_{0,1}, t_{0,2}]$ if $g(t_{0,1}) = g(t_{0,2})$. Thus we have located the minimum in an interval $[a_1, b_1]$ smaller than $[a_0, b_0]$ and we can proceed iteratively. The method would be most efficient if we only need to evaluate g at *one* new point each time—that is, if either $t_{1,1}$ or $t_{1,2}$ would equal whichever of $t_{0,1}$ and $t_{0,2}$ lies in (a_1, b_1); to allow this, we never choose $a_1 = t_{0,1}, b_1 = t_{0,2}$ but in that case of $g(t_{0,1}) = g(t_{0,2})$ we define $a_1 = a_0, b_1 = t_{0,2}$. If one seeks the smallest final interval $[a_m, b_m]$ for a given m, then it is known [Kiefer (1957), Spang (1962)] that one should choose

$$t_{i,1} = \frac{F_{m-1-i}}{F_{m+1-i}}(b_i - a_i) + a_i, \qquad t_{i,2} = \frac{F_{m-i}}{F_{m+1-i}}(b_i - a_i) + a_i$$

where $F_0 = 1, F_1 = 1, F_j = F_{j-1} + F_{j-2}$ are the *Fibonacci numbers*. This *Fibonacci search* always requires only one evaluation of g per step. On the final step, one takes

$$\begin{aligned} t_{m-1,1} &= (\tfrac{1}{2} + \epsilon)(b_{m-1} - a_{m-1}) + a_{m-1} \\ t_{m-1,2} &= \tfrac{1}{2}(b_{m-1} - a_{m-1}) + a_{m-1} \end{aligned}$$

in order to isolate the minimum best. The final interval has width

$$b_m - a_m = \frac{(b_0 - a_0)}{2F_m} + \epsilon$$

Since $F_{20} > 10^4$, we see that the intervals shrink rapidly. It is known that for large i, we have

$$\frac{F_{i-1}}{F_{i+1}} \sim 0.382, \quad \frac{F_i}{F_{i+1}} \sim 0.618$$

which allows one to use the simpler formulas:

$$t_{i,1} = 0.382(b_i - a_i) + a_i$$
$$t_{i,2} = 0.618(b_i - a_i) + a_i$$

The final interval in this way satisfies

$$b_m - a_m = (0.618)^m(b_0 - a_0)$$

Thus one can isolate the minimum in this way as accurately as desired.

Next we turn to methods using interpolation, although some of our remarks apply to direct-search techniques as well. Some of the procedures first seek an interval in which the minimizing point t^* lies. Usually this is done by taking some number t_e as an estimate of t^* and then evaluating g at $0, t_e, \alpha_2 t_e, \alpha_3 t_e, \ldots$, for some sequence α_i (often $\alpha_i = 2^i$) and stopping at the first instance that the values of g do not decrease; if one is willing to evaluate $g'(\alpha_i t_e)$ as well, one can also stop whenever $g'(\alpha_i t_e)$ becomes positive. If the termination procedure occurs at $t = t_e$, then t_e is reduced and the process restarted. Thus we finally find $\alpha_i t_e$ with $g(\alpha_i t_e) < g(\alpha_{i-1} t_e)$, $g(\alpha_i t_e) < g(\alpha_{i+1} t_e)$ and t^* is isolated in $[\alpha_{i-1} t_e, \alpha_{i+1} t_e]$. The number of evaluations of g will be reduced if t_e, at least near the solution x^*, is a good estimate, for then one would expect to isolate t^* in $[0, \alpha_2 t_e]$ every time. In fact if near the soution x^* one sets $t'_e = \frac{1}{2} t_e$ where t_e is asymptotically correct, then we should isolate t^* easily in $[t'_e, 3t'_e]$ and a choice of $t^* = t'_e$ or $2t'_e$, whichever gave the lower f-value, would lead to convergence, as we saw at the start of this section. In this light we see that Theorem 5.8.1 on the convergence of the conjugate-gradient method with c_n determined as

$$c_n = \frac{\langle r_n, p_n \rangle}{\langle J'_n p_n, p_n \rangle}$$

can be considered as providing a good estimate t_e which is asymptotically the correct t^*; this has been used [Daniel (1967a)] as t_e and has given good results. If $\{p_n\}$ is any admissible sequence of directions and the functional f on $\mathbb{R}^l$ satisfies $0 < aI \leq f''_x \leq AI$, the analogous choice for t_e is

$$t_e = \frac{\langle -\nabla f(x_n), p_n \rangle}{\langle f''_{x_n} p_n, p_n \rangle}$$

It has been shown [Elkin (1968)] that one obtains global convergence with $t_n = \beta_n t_e$ where $0 < \epsilon \leq \beta_n \leq 2 - \epsilon$, and of course that t_e is asymptotically correct. Thus linearization can always be used to get a good estimate t_e if one can afford to evaluate f''_x. If one cannot compute f''_x but has an estimate f^* for the minimum value of f, then

$$t'_e = \frac{f(x_n) - f^*}{\langle -\nabla f(x_n), p_n \rangle}$$

is usually an underestimate of t^* near the solution x^* while $2t'_e$ is usually an overestimate near the solution. Another choice of the estimate t_e is simply the value of the actual step used the preceding time; in the end this usually requires little computation to trap t^* in an interval.

Now we turn to the problem of locating t^* more accurately by interpolation. The interpolation procedures are sometimes used without first bounding t^*; in this case, the "interpolation" becomes extrapolation but the formulas are essentially the same. Such methods are, therefore, contained within the ensuing discussion, although we generally prefer the methods which first bound t^*.

Nearly all of the analysis of minimization methods is based on the assumption that the function f is nearly quadratic near its minimizing point; thus it would be reasonable, and asymptotically exact, to approximate $f(x_n + tp_n)$ by a quadratic or $\langle \nabla f(x_n + tp_n), p_n \rangle$ by a linear function. Although it is often assumed in the literature that ∇f is much more costly to evaluate than is f, in certain important practical problems such as arise from differential equations we can afford to evaluate ∇f reasonably often. In this case, one can then try to solve $\langle \nabla f(x_n + tp_n), p_n \rangle = 0$, and it is quite reasonable to use linear interpolation to solve this—that is, use the secant method. Since one can only approximately satisfy the equation, scaling becomes important and one should probably treat instead

$$\left\langle \frac{\nabla f(x_n + tp_n)}{\|\nabla f(x_n + tp_n)\|}, \frac{p_n}{\|p_n\|} \right\rangle = 0$$

It is no longer clear that linear interpolation is appropriate here, so one might consider using quadratics—that is, Muller's method. In our experience, however, the linear interpolation is usually satisfactory. Essentially the same idea as using a quadratic on the gradient equation is that of using a cubic on the function f. Again we assume that we can evaluate ∇f conveniently. Thus we suppose that we have the real-valued function $g(t)$ of one real variable to minimize, and that we know the function values g_1, g_2 and the derivatives g'_1, g'_2 at two points $\tau_1 < \tau_2$; we wish to interpolate the data by a cubic and then minimize the cubic. This method is usually used when t^* is bracketed by $[\tau_1, \tau_2]$, in which case the next estimate is a zero of the quadratic derivative in

$[\tau_1, \tau_2]$. In many implementations for which the basic interval $[\tau_1, \tau_2]$ was chosen so that estimating t^* by τ_1 would yield convergence to x^*, the interpolation is performed only until an estimate of t^* is provided by the scheme at which g is smaller than at τ_1 or τ_2. This guarantees very accurate minimization near x^*.

Quadratic interpolation to the values of g at three points followed by minimization of the interpolating quadratic appears in general to be an excellent scheme, particularly if evaluation of ∇f is very costly. This is the method commonly used with algorithms which never evaluate ∇f(see Chapter 9). A variation is to use one value of the derivative for at least the first estimation of t^*; this is easy, since the derivative at $t=0$—that is, at x—is often known.

EXERCISE. Write an algorithm implementing one of the above interpolation schemes.

We are not aware of any good tests comparing the efficiencies of the various interpolation methods; our limited experience indicates that the simpler approaches—say, using quadratics on f-values—are usually satisfactory. One should not generally spend much effort locating t^* unless one appears to be very near x^* and wants to avoid badly over- or undershooting it. Even then, driving the cosine of the angle between $\nabla f(x_{n+1})$ and p_n to be less than 0.1 often is quite satisfactory. Precisely how accurate one needs be at this point depends on the criterion used for determining convergence of $\{x_n\}$ to x^*; if this is based on the size of $\| x_{n+1} - x_n \|$, clearly one must approximate t^*p_n to at least the accuracy demanded for the cutoff of $\| x_{n+1} - x_n \|$. As is always true with numerical methods, no single, good, universally applicable method is known for deciding when $\{x_n\}$ has converged. If necessary, one can use the very stringent test of moving away from the computed "x^*" and starting the algorithm over to see if the sequence returns toward "x^*" again. No really good method is known.

General references: Fletcher (1965, 1968), Fletcher–Powell (1963), Fletcher–Reeves (1964), Kowalik–Osborne (1968), Powell (1964a, b), Stewart (1967).

7 VARIABLE-METRIC GRADIENT METHODS IN $\mathbb{R}^l$

7.1. INTRODUCTION

The earliest gradient-type methods relied strictly on the steepest-descent direction; that is, to minimize f given an initial point x_0, one wrote

$$f(x_0 + tp) = f(x_0) + t\langle \nabla f(x_0), p\rangle + o(t)$$

so that

$$\frac{d}{dt} f(x_0 + tp)\bigg|_{t=0} = \langle \nabla f(x_0), p\rangle$$

which is the most negative for directions p with $||p|| = 1$ if

$$p = \frac{-\nabla f(x_0)}{||\nabla f(x_0)||}$$

The results of Chapters 4, 5, and 6 show, however, that using the steepest-descent direction itself is not necessary; essentially, any direction bounded away from being orthogonal to $-\nabla f(x_0)$ will suffice. In fact, as we saw with respect to the conjugate-gradient (CG) methods in Chapter 5, one may well obtain remarkably more rapid convergence by purposefully avoiding the steepest-descent direction. Let us therefore consider other ways of generating directions.

7.2. VARIABLE-METRIC DIRECTIONS

Consider again the expression

$$f(x_0 + tp) = f(x_0) + t\langle \nabla f(x_0), p\rangle + o(t)$$

Suppose Q_0 is some self-adjoint positive-definite operator ($l \times l$ matrix) in $\mathbb{R}^l$. Then we can write

$$\begin{aligned} f(x_0 + tp) &= f(x_0) + t\langle \nabla f(x_0), p \rangle + o(t) \\ &= f(x_0) + t\langle Q_0^{-1} \nabla f(x_0), Q_0 p \rangle + o(t) \end{aligned} \tag{7.2.1}$$

Since Q_0 is self-adjoint and positive-definite, we can use it to define a new *metric*—that is, a new inner product—on $\mathbb{R}^l$ which will determine a topology equivalent to the usual one; precisely, we define the inner product $[\cdot, \cdot]$ via

$$[x, y] \equiv \langle x, Q_0 y \rangle$$

Then we can rewrite Equation 7.2.1 as

$$f(x_0 + tp) = f(x_0) + t[Q_0^{-1} \nabla f(x_0), p] + o(t)$$

and suddenly the steepest-descent direction with respect to this new metric has become $-Q_0^{-1} \nabla f(x_0)$. Since Q_0^{-1} is itself positive-definite and self-adjoint, we call the direction $-H_0 \nabla f(x_0)$ where $H_0 = Q_0^{-1}$ is positive-definite and self-adjoint. If we use a different "H" (that is, "Q") at each successive approximation x_n to the minimizing point x^*—that is, if we use a different metric each time—we thereby generate the sequence of directions

$$p_n \equiv -H_n \nabla f(x_n)$$

A method of this type is called a *variable-metric method* [Davidon (1959, 1968), Fletcher–Powell (1963)].

From this viewpoint it is clear that any method yielding directions p_n such that $\langle p_n, -\nabla f(x_n) \rangle > 0$ if $\nabla f(x_n) \neq 0$ is a variable-metric method, since one can always find a self-adjoint positive-definite operator H_n such that $p_n = -H_n \nabla f(x_n)$; the purpose of viewing gradient methods in this fashion, however, is to discover what properties H_n should have to generate directions that are good ones. We have already seen in Section 5.5, for example, that if one is trying to solve $Mx = k$ where M is self-adjoint and positive-definite and if one chooses H_n as the orthogonal-projection operator, with respect to the inner product $\langle x, My \rangle$, onto the subspace orthogonal in the $\langle x, My \rangle$ sense to $p_0, p_1, \ldots, p_{n-1}$, then the method obtained is the conjugate-gradient method. As we shall later see, the Davidon method is in fact a way of generating the conjugate-gradient iterates directly from a recursively defined set of matrices $\{H_n\}$. For more general problems with $J(x) \equiv \nabla f(x)$ nonlinear, H_n becomes the orthogonal projection in the $\langle x, J_n' y \rangle$ sense onto the subspace $\langle \cdot, J_n' \cdot \rangle$-orthogonal to p_{n-1} and—asymptotically, of course—to $p_0, \ldots, p_{n-2}$ also. Thus we can consider that the power of the conjugate-gradient method compared to the steepest-descent method comes from the former's use of a good variable metric. In Yakolev (1965), gradient-type methods are considered strictly in the setting of variable-metric methods—that is,

$$x_{n+1} = x_n - t_n H_n \nabla f(x_n)$$

for some sequence of operators H_n and steps t_n. Most of the results there concern convergence under various choices of t_n given certain properties of H_n such as

$$0 < a\langle p, H_n^{-1}p\rangle \leq \langle p, f_x'' p\rangle \leq A\langle p, H_n^{-1}p\rangle$$

These correspond, with some minor changes, to the methods of Chapter 4, although more detailed convergence rates are often given in Yakovlev (1965). Thus we consider the methods in this completely general setting no further.

In a sense the best metric would be one which turns the level curves $f(x) = c$ into spheres so that the interior normal direction to the surface—that is, $-\nabla f(x)$—points to the point minimizing f. For quadratic functionals

$$f(x) = \langle h - x, M(h - x)\rangle = [h - x, h - x]$$

where

$$[u, v] = \langle u, Mv\rangle$$

the new metric defined by $[\cdot, \cdot]$ makes the level curves appear to be spheres; this leads to the direction $-M^{-1}\nabla f(x) = 2(h - x)$—that is, directly toward the solution. Analogously, for nonlinear equations, the optimum metric would appear to be given by $\langle \cdot, J_n' \cdot\rangle$ and thus generates the direction

$$p_n = -J_n'^{-1}J(x_n)$$

This is the direction of *Newton's* method. Because of this intuitive viewpoint and because Newton's method leads to quadratic convergence [Kantorovich–Akilov (1964), Rall (1969)], one often tries to pick the variable-metric formulation to mimic Newton's method; thus variable-metric methods are also called *quasi-Newton methods* [Broyden (1965, 1967), Zeleznik (1968)]. Because of the situation in the constrained case (see Section 4.10), one might not greatly expect quadratic convergence from mimicking the Newton process if one proceeds close to the Newton direction to the minimum of f along that line rather than using the pure Newton step

$$x_{n+1} = x_n - J_n'^{-1}J(x_n)$$

However, the value of t_n which minimizes $f(x_n + tp_n)$ is asymptotically

$$\hat{t}_n = \frac{\langle r_n, p_n\rangle}{\langle p_n, J_n' p_n\rangle} = \frac{\langle r_n, J_n'^{-1} r_n\rangle}{\langle r_n, J_n'^{-1} r_n\rangle} = 1$$

in this case, and thus near the solution x^* the minimization along $x_n + tp_n$ nearly leads to the normal Newton step. While one should then hope for quadratic convergence, most results known to us guarantee only superlinear convergence [Levitin–Poljak (1966a), Yakovlev (1965)]. From what we have done, this can most easily be seen from the viewpoint of conjugate gradients. In Sections 5.3, 5.4, and 5.5 we considered a very general form of conjugate-gradient methods involving arbitrary self-adjoint positive-definite operators H and K, while in Section 5.6 such extra operators were missing. Clearly one may define a general method using operators H_x, K_x at each point x and develop convergence theory and error estimates in terms of the operator $T_x = K_x J_x'^* H_x J_x'$ just as in the quadratic-functional case; this is done in Daniel (1965, 1967a, b), and the convergence rates are given via the spectral bounds a, A of T_x as usual. If one takes $H_x = K_x = J_x'^{-1}$, where J_x' is self-adjoint, uniformly positive-definite, and uniformly bounded, one gets $T_x = I$ and $a = A = 1$, which implies superlinear convergence. In this case, of course, $p_n = J_n'^{-1} r_n = -J_n'^{-1} J(x_n)$ and we have the minimization modification of Newton's method and a proof of superlinear convergence. It is possible, however, to show that the convergence is actually quadratic. If we let

$$x_n' = x_n - J_n'^{-1} J(x_n)$$

and pick x_{n+1} so that $f(x_{n+1}) \leq f(x_n')$, from $0 < aI \leq J_x' \leq AI$ for scalars a, A, one can conclude

$$\begin{aligned}\frac{a}{2} \|x_{n+1} - x^*\|^2 &\leq f(x_{n+1}) - f(x^*) \leq f(x_n') - f(x^*)\\ &\leq \frac{A}{2} \|x_n' - x^*\|^2\\ &\leq \text{const} \times \|x_n - x^*\|^4\end{aligned}$$

so that

$$\|x_{n+1} - x^*\| \leq \text{const} \times \|x_n - x^*\|^2$$

EXERCISE. Provide the details for the above argument showing $\|x_{n+1} - x^*\| \leq \text{const} \times \|x_n - x^*\|^2$.

Thus we hope that a good quasi-Newton or variable-metric method will yield very rapid convergence. To obtain this convergence, most of the methods always choose t_n by minimization along $x_n + tp_n$; in some cases this is necessary in order that the next "metric"—that is, H_{n+1}, defined often in terms of x_{n+1}—be a good one. A recent method of Davidon (1968), however, attempts to pick t_n automatically and include it in H_n so that really we have $t_n = 1$; this can be viewed as one of the interval methods along $x_n + tp_n$. The iteration is as follows:

Given x_n and H_n, one sets $x'_n = x_n - H_n \nabla f(x_n)$. Set

$$\rho_n = \langle \nabla f(x'_n), H_n \nabla f(x'_n) \rangle, \qquad \gamma_n = -\frac{\langle \nabla f(x_n), H_n \nabla f(x'_n) \rangle}{\rho_n}$$

For two fixed positive constants α, β, with $0 < \alpha < 1 < \beta$, one then defines

$$\lambda_n = \begin{cases} \alpha \quad \text{if} \quad \dfrac{-\alpha}{1+\alpha} \leq \gamma_n < \dfrac{\alpha}{1-\alpha} \\ -\dfrac{\gamma_n}{1+\gamma_n} \quad \text{if} \quad \dfrac{-\beta}{1+\beta} \leq \gamma_n < \dfrac{-\alpha}{1+\alpha} \\ \beta \quad \text{if} \quad \dfrac{-\beta}{\beta - 1} \leq \gamma_n < \dfrac{-\beta}{\beta+1} \\ \dfrac{\gamma_n}{1-\gamma_n} \quad \text{otherwise} \end{cases}$$

and defines

$$H_{n+1} = H_n + \frac{(\lambda_n - 1) H_n \nabla f(x'_n) [H_n \nabla f(x'_n)]^*}{\rho}$$

where * denotes conjugate transpose. If $f(x'_n) \geq f(x_n)$, then we set $x_{n+1} = x_n$; if $f(x'_n) < f(x_n)$, then we set $x_{n+1} = x'_n$.

The value of λ_n that is chosen minimizes the length of

$$H_{n+1}[\nabla f(x'_n) - \nabla f(x_n)] - (x'_n - x_n)$$

in the inner product $\langle \cdot, H_{n+1}^{-1} \cdot \rangle$; this distance would be zero for Newton's method applied to a quadratic f, and therein lies the reason for so choosing λ_n. It can be shown [Davidon (1968)] that if H_0 is positive-definite, then each H_n is positive-definite. If f is a quadratic,

$$f(x) = \langle h - x, M(h - x) \rangle$$

and if

$$\frac{\gamma_n}{1 + \gamma_n} \in [\alpha, \beta]$$

for all n, then $x_n \longrightarrow h$ and, in fact, in $\mathbb{R}^l$, $x_l = h$. The only hypothesis known to guarantee that $\gamma_n/(1 + \gamma_n) \in [\alpha, \beta]$ is as follows [Vercoustre (1969)]: if

$$0 < \alpha \langle x, H_0 x \rangle \leq \langle x, Mx \rangle \leq \beta \langle x, H_0 x \rangle \quad \text{and} \quad \beta = \frac{1}{\alpha}$$

then

$$\frac{\gamma_n}{1+\gamma_n} \in [\alpha, \beta]$$

Thus, even for quadratics, one cannot guarantee convergence in general. In fact, if $\gamma_n = -\frac{1}{2}$, then $\lambda_n = 1$ and $H_{n+1} = H_n$; so if $\gamma_n = -\frac{1}{2}$ and $f(x_n') \geq f(x_n)$, the iteration halts at x_n. Computationally, a similar phenomenon has been observed, and the method as presently developed does not appear to this author to be exceptionally useful; when the method *does* work, it works fairly well [Vercoustre (1969)].

Let us return to the question of generating the matrices H_n. If we used precisely Newton's method with $H_n = f_n''^{-1}$, we would have

$$\begin{aligned} H_n[\nabla f(x_n) - \nabla f(x_{n-1})] &= f_n''^{-1}[f_n''(x_n - x_{n-1}) + o(\|x_n - x_{n-1}\|)] \\ &= x_n - x_{n-1} + o(\|x_n - x_{n-1}\|) \end{aligned}$$

so it seems reasonable to ask that in general

$$H_n[\nabla f(x_n) - \nabla f(x_{n-1})] = x_n - x_{n-1}$$

If we let

$$H_{n+1} = H_n + B_n$$

then we have

$$\begin{aligned} H_{n+1}[\nabla f(x_{n+1}) - \nabla f(x_n)] &= H_n[\nabla f(x_{n+1}) - \nabla f(x_n)] + B_n[\nabla f(x_{n+1}) - \nabla f(x_n)] \\ &= x_{n+1} - x_n \end{aligned}$$

For convenience, we define

$$\delta_n = \nabla f(x_{n+1}) - \nabla f(x_n)$$

and then we must pick B_n so that

$$B_n\delta_n = x_{n+1} - x_n - H_n\delta_n$$

Computationally, we desire B_n to be rather simple; for example, one might allow B_n two degrees of freedom and set

$$B_n = (x_{n+1} - x_n)q_n^* - H_n\delta_n z_n^*$$

where (7.2.2)

$$\langle q_n, \delta_n \rangle = \langle z_n, \delta_n \rangle = 1$$

and * denotes conjugate transpose. This defines a very general class of variable-metric methods in terms of the two families $\{q_n\}$ and $\{z_n\}$ [Broyden (1967)].

If each H_n is positive-definite with

$$0 < a_n I \leq H_n \leq A_n I \quad \text{and} \quad \frac{a_n}{A_n} \geq \epsilon > 0$$

then

$$\begin{aligned}\left\langle -\nabla f(x_n), \frac{p_n}{\|p_n\|} \right\rangle &= \frac{\langle \nabla f(x_n), H_n \nabla f(x_n) \rangle}{\| H_n \nabla f(x_n) \|} \\ &\geq \frac{a_n \|\nabla f(x_n)\|^2}{A_n \|\nabla f(x_n)\|} \\ &\geq \epsilon \|\nabla f(x_n)\|\end{aligned}$$

and we see that the direction sequence is admissible and that the variable-metric method will yield convergence under these conditions. Using the approach of Remark 1 after Theorem 4.2.4, for the case in which ∇f is Lipschitzian, we would only require

$$\sum_{n=0}^{\infty} \left(\frac{a_n}{A_n}\right)^2 = \infty$$

for convergence. Rather than try to see how one should pick H_n so as to satisfy either of these conditions, we note that we were seeking not just a convergent method but one with Newton-like rapid convergence. For the conjugate-gradient method we saw that the $1/l$-superlinear convergence in $\mathbb{R}^l$ depends on the fact that the method exactly minimizes quadratics in finitely many steps. We consider other such algorithms in the following sections.

7.3. EXACT METHODS FOR QUADRATICS

It has been widely stated that any method which minimizes a quadratic exactly in a finite number of steps will be a good one for more general functionals and should exhibit superlinear or at least very rapid convergence; the $1/l$-superlinear convergence of the conjugate-gradient method depends largely on its exactness for quadratics in $\mathbb{R}^l$. This by itself is certainly not sufficient in fact to define a good method; one at least should require the direction sequence to be admissible. Exactness does, however, appear to be a reasonably useful property which we should try to obtain for the variable-metric methods.

EXERCISE. Find an exact method which is useless on general functionals.

Necessary and sufficient conditions for a variable-metric method as presented here to be exact—that is, exact for quadratics in $\mathbb{R}^l$—do not appear to be known. As our remarks prior to Theorem 5.2.2 indicate, however, it is sufficient and "almost necessary" that the directions be generated by conjugate directions; experience also seems to indicate that the conjugate-direction methods of some form are usually the best. This in fact has been the approach used implicitly, if not explicitly, in the presentation of most methods. One way of developing the methods is as follows [Broyden (1967), Zeleznik (1968)]:

Since for quadratics with

$$f(x) = \tfrac{1}{2}\langle h - x, M(h - x)\rangle$$

we always have

$$M^{-1}\delta_n \equiv M^{-1}[\nabla f(x_{n+1}) - \nabla f(x_n)] = x_{n+1} - x_n$$

and we are trying to duplicate Newton's method, let us insist that H_{n+1} not only satisfy $H_{n+1}\delta_n = x_{n+1} - x_n$ but also satisfy

$$H_{n+1}\delta_i = x_{i+1} - x_i, \qquad i = 0, 1, \ldots, n$$

This gives, for $i \leq n - 1$,

$$x_{i+1} - x_i = H_{n+1}\delta_i = H_n\delta_i + B_n\delta_i = x_{i+1} - x_i + B_n\delta_i$$

and hence

$$B_n\delta_i = 0 \quad \text{for} \quad i = 0, 1, \ldots, n - 1$$

In the special case of Equation 7.2.2, where we allow two degrees of freedom, we find

$$B_n = (x_{n+1} - x_n)q_n^* - H_n\delta_n z_n^*$$

where (7.3.1)

$$\langle q_n, \delta_i\rangle = \langle z_n, \delta_i\rangle = \begin{cases} 1 & \text{if} \quad i = n \\ 0 & \text{if} \quad i \leq n - 1 \end{cases}$$

Such q_n and z_n exist if δ_n is linearly independent of the set $\{\delta_0, \delta_1, \ldots, \delta_{n-1}\}$. If this is true for all n, then $\delta_0, \ldots, \delta_{n-1}$ are linearly independent and we have

$$MH_{n+1}\delta_i = M(x_{i+1} - x_i) = \delta_i, \qquad i = 0, 1, \ldots, n$$

Thus in $\mathbb{R}^l$,

$$MH_l\delta_i = \delta_i, \qquad i = 0, 1, \ldots, l-1$$

which says that MH_l has l linearly independent eigenvectors δ_i associated with the eigenvalue 1 and hence

$$MH_l = I$$

Thus $p_l = -H_l\nabla f(x_l)$ will yield the exact solution $x_{l+1} = h$ for quadratic f.

Since the argument above did not depend on the two-parameter nature of the matrices H_n, we have proved the following theorem.

THEOREM 7.3.1. Suppose that $\{\delta_0, \ldots, \delta_n\}$ is a linearly independent set of vectors for $0 \leq n \leq l-1$ in $\mathbb{R}^l$, where $\delta_i \equiv M(x_{i+1} - x_i)$, and $H_n\delta_i = x_{i+1} - x_i$ for $0 \leq i \leq n-1$; then $x_{l+1} = h$, the solution, for quadratic $f(x) = \frac{1}{2}\langle x - h, M(x-h)\rangle$.

COROLLARY 7.3.1 [Vercoustre (1969)]. Suppose that $H_n\delta_i = x_{i+1} - x_i$ for $0 \leq i \leq n-1$, $0 \leq n \leq l-1$, where $\delta_i = M(x_{i+1} - x_i)$; suppose that $H_n\delta_n \neq x_{n+1} - x_n$ for $0 \leq n \leq l$. Then $x_{l+1} = h$, the solution for quadratics $f(x) = \frac{1}{2}\langle x - h, M(x-h)\rangle$.

Proof: If for some n we have $\delta_n = \sum_{i=0}^{n-1} \tau_i\delta_i$, then

$$\begin{aligned} H_n\delta_n &= \sum_{i=0}^{n-1} \tau_i H_n\delta_n = \sum_{i=0}^{n-1} \tau_i(x_{i+1} - x_i) = \sum_{i=0}^{n-1} \tau_i M^{-1}\delta_i \\ &= M^{-1}\sum_{i=0}^{n-1} \tau_i\delta_i = M^{-1}\delta_n = x_{n+1} - x_n \end{aligned}$$

which is a contradiction. Thus $\{\delta_0, \ldots, \delta_n\}$ is linearly independent for all n. Q.E.D.

We now suppose that H_{n+1} is symmetric and that t_n is always chosen to minimize $f(x_n + tp_n)$, so that

$$\begin{aligned} 0 = \langle \nabla f(x_{n+1}), t_np_n\rangle &= \langle \nabla f(x_{n+1}), H_{n+1}\delta_n\rangle \\ &= \langle H_{n+1}\nabla f(x_{n+1}), \delta_n\rangle \\ &= -\langle p_{n+1}, \delta_n\rangle, \quad \text{for} \quad n = 0, 1, \ldots, r \end{aligned}$$

Under these hypotheses for the two-parameter methods of Equation 7.3.1 it can then be proved [Broyden (1967)] that $\langle p_i, Mp_j\rangle = 0$ if $i \neq j$ for $0 \leq i$, $j \leq r$ and, therefore, that we have a conjugate-direction method. The proof goes roughly as follows. From the definitions of H_{n+1} and p_{n+1}, it easily follows that

$$p_{n+1} = \alpha_np_n + \beta_nH_n\delta_n \quad \text{for some scalars} \quad \alpha_n, \beta_n$$

Now,

$$\langle p_{n+1}, Mp_n\rangle = \frac{1}{t_n}\langle p_{n+1}, M(x_{n+1} - x_n)\rangle = \frac{1}{t_n}\langle p_{n+1}, \delta_n\rangle = 0$$

for

$$n = 0, 1, \ldots, r$$

We then get

$$\begin{aligned}\langle p_{n+1}, Mp_{n-1}\rangle &= \langle \alpha_n p_n + \beta_n H_n \delta_n, Mp_{n-1}\rangle \\ &= \alpha_n \langle p_n, Mp_{n-1}\rangle + \beta_n \langle H_n\delta_n, Mp_{n-1}\rangle \\ &= \beta_n \left\langle H_n\delta_n, \frac{\delta_{n-1}}{t_{n-1}}\right\rangle = \frac{\beta_n}{t_{n-1}}\langle \delta_n, H_n\delta_{n-1}\rangle \\ &= \frac{\beta_n}{t_{n-1}}\langle \delta_n, t_{n-1}p_{n-1}\rangle = 0, \qquad n = 0, 1, \ldots, r\end{aligned}$$

The induction then proceeds easily to give

$$\langle p_{n+1}, Mp_i\rangle = 0, \qquad i = 0, 1, \ldots, n, \qquad n = 0, 1, \ldots, r$$

Thus we have found a two-parameter class of exact variable metric methods; the admissibility of the direction sequence for nonquadratic functionals still remains unknown, however. Since a study of the admissibility requires considerable specialization of the vectors q_n, z_n, we consider this question for special methods, although little information is available even in special cases.

7.4. SOME PARTICULAR METHODS

We consider first the class of variable-metric methods [Broyden (1967)] defined via

$$\begin{aligned} q_n &\equiv (\alpha_n p_n - \beta_n H_n \delta_n) \\ z_n &\equiv (\gamma_n H_n \delta_n + \beta_n t_n p_n) \\ \alpha_n &= \frac{(1 + \beta_n \langle \delta_n, H_n \delta_n\rangle)}{\langle \delta_n, p_n\rangle} \\ \gamma_n &= \frac{(1 - \beta_n t_n \langle \delta_n, p_n\rangle)}{\langle \delta_n, H_n\delta_n\rangle} \end{aligned} \tag{7.4.1}$$

where β_n is arbitrary, t_n is chosen so $\langle \nabla f(x_{n+1}), p_n\rangle = 0$, and H_0 is symmetric.

By a straightforward inductive argument [Broyden (1967)] or by using Corollary 7.3.1, one can show that, if M is symmetric and nonsingular, then the δ_i are linearly independent and hence the method is exact.

THEOREM 7.4.1. If H_0 and M are positive-definite and $\beta_n \geq 0$, then it follows that H_n is positive-definite for each n.

Proof: The proof goes by induction. If H_n is positive-definite, let $LL^* = H_n$. If we let $u = L^*\nabla f(x_n)$, $v = L^*x$, and $w = L^*\delta_n$, we then have

$$\langle x, H_{n+1}x\rangle = \langle v, v\rangle - \frac{\langle v, w\rangle^2}{\langle w, w\rangle} + \frac{t_n\langle u, v\rangle^2}{\langle u, u\rangle} + \frac{\beta_n t_n}{\langle u, u\rangle\langle w, w\rangle}[\langle v, u\rangle\langle v, w\rangle + \langle w, w\rangle\langle v, u\rangle]^2$$

The only possible negative terms come from

$$\langle v, v\rangle \qquad \frac{\langle v, w\rangle^2}{\langle w, w\rangle}$$

which in fact is nonnegative by the Schwarz inequality and is positive unless $v = \lambda w$. If the term $\langle u, v\rangle^2$ is also zero, then $\langle u, w\rangle = 0$ but $\langle u, w\rangle = -\langle u, u\rangle \neq 0$. Therefore, $\langle x, H_{n+1}x\rangle > 0$ if $x \neq 0$ and hence H_{n+1} is positive-definite. Q.E.D.

Since only finitely many iterations need be used for quadratics, we have

$$a_n I \leq H_n \leq A_n I \quad \text{with} \quad \frac{a_n}{A_n} \geq \epsilon > 0$$

in that case. If we use the algorithm for more general functionals, however, we cannot immediately conclude such bounds (see, however, Theorems 7.4.2 and 7.4.5). Similarly, it is not known whether or not such bounds exist for quadratics in infinite-dimensional spaces, a result which would at least give some indications for the nonquadratic case in $\mathbb{R}^l$. At this time numerical experience testing various choices of the parameters β_n is rather limited; thus the method for arbitrary β_n requires further study both theoretically and computationally. In practice, when one uses this method one seldom actually performs an exact minimization along the direction p_n to reach x_{n+1}—that is, one seldom has $\langle \nabla f(x_{n+1}), p_n\rangle = 0$; it is striking to note, however, that if $\langle \nabla f(x_{n+1}), p_n\rangle = 0$ for all n, then the directions $p_n/\|p_n\|$ are *independent* of the parameters $\{\beta_i\}$. Computationally, however, one finds great dependence on the choice of these parameters.

THEOREM 7.4.2. Under the assumptions of this section, if $\langle \nabla f(x_{n+1}), p_n\rangle = 0$, then the direction

$$\frac{p_{n+1}}{\|p_{n+1}\|}$$

determined by Equation 7.4.1 is independent of β_n.

Proof:

$$\begin{aligned}-p_{n+1} &= H_{n+1}\nabla f(x_{n+1}) \\ &= H_n\nabla f(x_{n+1}) - H_n\delta_n\frac{\langle\delta_n, H_n\nabla f(x_n)\rangle}{\langle\delta_n, H_n\delta_n\rangle} \\ &\quad + t_n p_n\frac{\langle p_n, \nabla f(x_{n+1})\rangle}{\langle p_n, \delta_n\rangle} + \beta_n\Gamma_n\end{aligned}$$

where

$$\begin{aligned}\Gamma_n \equiv t_n p_n&\left\{\frac{\langle\delta_n, H_n\delta_n\rangle}{\langle p_n, \delta_n\rangle}\langle p_n, \nabla f(x_{n+1})\rangle - \langle\delta_n, H_n\nabla f(x_{n+1})\rangle\right\} \\ &- H_n\delta_n\left\{t_n\langle p_n, \nabla f(x_{n+1})\rangle - t_n\langle p, \delta_n\rangle\frac{\langle\delta_n, H_n\nabla f(x_{n+1})\rangle}{\langle\delta_n, H_n\delta_n\rangle}\right\}\end{aligned}$$

Using $\langle p_n, \nabla f(x_{n+1})\rangle = 0$, we have

$$\begin{aligned}-p_{n+1} = H_n\nabla f(x_{n+1}) &- H_n\delta_n\frac{\langle\delta_n, H_n\nabla f(x_{n+1})\rangle}{\langle\delta_n, H_n\delta_n\rangle} \\ &+ \beta_n t_n\left\{p_n\langle-\delta_n, H_n\nabla f(x_{n+1})\rangle + H_n\delta_n\frac{\langle p_n, \delta_n\rangle\langle\delta_n, H_n\nabla f(x_{n+1})\rangle}{\langle\delta_n, H_n\delta_n\rangle}\right\}\end{aligned}$$

Now,

$$\frac{1}{\langle\delta_n, H_n\nabla f(x_{n+1})\rangle}$$

times the term in braces { } in the above expression for $-p_{n+1}$ yields

$$\begin{aligned}\frac{\{\cdot\}}{\langle\delta_n, H_n\nabla f(x_{n+1})\rangle} &= -p_n + \frac{\langle p_n, \delta_n\rangle}{\langle\delta_n, H_n\delta_n\rangle}H_n\delta_n \\ &= -p_n - \frac{\langle\delta_n, H_n\nabla f(x_n)\rangle}{\langle\delta_n, H_n\delta_n\rangle}H_n\delta_n \\ &= -p_n - \frac{\langle\delta_n, H_n(-\delta_n)\rangle + \langle\delta_n, H_n\nabla f(x_{n+1})\rangle}{\langle\delta_n, H_n\delta_n\rangle}H_n\delta_n \\ &= H_n\nabla f(x_n) + H_n\delta_n - \frac{\langle\delta_n, H_n\nabla f(x_{n+1})\rangle}{\langle\delta_n, H_n\delta_n\rangle}H_n\delta_n \\ &= H_n\nabla f(x_{n+1}) - \frac{\langle\delta_n, H_n\nabla f(x_{n+1})\rangle}{\langle\delta_n, H_n\delta_n\rangle}H_n\delta_n\end{aligned}$$

which is the leading part in the expression for $-p_{n+1}$. Thus we have

$$\begin{aligned}p_{n+1} = &-[1 + \beta_n t_n\langle\delta_n, H_n\nabla f(x_{n+1})\rangle] \\ &\times\left\{H_n\nabla f(x_{n+1}) - H_n\delta_n\frac{\langle\delta_n, H_n\nabla f(x_{n+1})\rangle}{\langle\delta_n, H_n\delta_n\rangle}\right\}\end{aligned}$$

and $p_n/\|p_n\|$ is independent of β_n. Q.E.D.

One reason for feeling that this class of methods *deserves* study is that the special case with $\beta_n = 0$ for all n—to which the others are in a sense equivalent, as we have seen— has been shown to be one of the most powerful methods available at present; this is the original *Davidon method* [Davidon (1959)] as modified by Fletcher and Powell (1963). In this way, H_n is modified as follows:

$$H_{n+1} = H_n + B_n$$
$$B_n = \frac{t_n p_n p_n^*}{\langle \delta_n, p_n \rangle} - \frac{H_n \delta_n (H_n \delta_n)^*}{\langle \delta_n, H_n \delta_n \rangle}$$
$$\delta_n = \nabla f(x_{n+1}) - \nabla f(x_n), \qquad x_{n+1} = x_n + t_n p_n$$

Therefore, from the results for arbitrary $\beta_n \geq 0$, we know that for quadratics this yields an exact method—in fact, a conjugate-direction method. A somewhat startling fact is that for quadratics this yields precisely the same iterates as a form of the conjugate-*gradient* method; this was first noted for $H_0 = I$ [Myers (1968)], but is true in general.

THEOREM 7.4.3. Let $f(x) = \frac{1}{2}\langle h - x, M(h - x)\rangle$, let M be self-adjoint and positive-definite, and let H_0 be self-adjoint and positive-definite. For a given x_0, let the Davidon method be used to generate points $x_1, x_2, \ldots$ by directions $p_0, p_1, \ldots$. Starting with $x_0' = x_0$, let the general form of the conjugate-gradient method of Section 5.3 be used with, in the notation of that section, $K \equiv H_0$, $H \equiv M^{-1}$, $N \equiv M$, $T = H_0 M$, generating a sequence $x_1', x_2', \ldots$ by directions $p_0', p_1', \ldots$. Then $x_n = x_n'$ and $p_n = \lambda_n p_n'$ for scalars λ_n for all n.

Proof: Define $g_n = -\nabla f(x_n)$. Since

$$\langle g_{n+1}, H_n g_n \rangle = -\langle g_{n+1}, p_n \rangle = 0$$

we have

$$\langle \delta_n, H_n \delta_n \rangle = \langle g_{n+1}, H_n g_{n+1} \rangle + \langle g_n, H_n g_n \rangle$$

Therefore,

$$p_n = H_{n-1} g_n - \frac{[H_{n-1} \delta_{n-1} \langle g_n, H_{n-1} g_n \rangle]}{\langle g_n, H_{n-1} g_n \rangle + \langle g_{n-1}, H_{n-1} g_n \rangle}$$

Then for $n < m$, we have

$$\begin{aligned} \langle g_m, H_{n-1} g_n \rangle &= \langle g_m, p_n \rangle + \langle g_m, H_{n-1}[g_{n-1} - g_n] \rangle \\ &\quad \times \frac{\langle g_n, H_{n-1} g_n \rangle}{\langle g_n, H_{n-1} g_n \rangle + \langle g_{n-1}, H_{n-1} g_{n-1} \rangle} \\ &= 0 + \langle g_m, p_n \rangle - \langle g_m, H_{n-1} g_n \rangle \\ &\quad \times \frac{\langle g_n, H_{n-1} g_n \rangle}{\langle g_n, H_{n-1} g_n \rangle + \langle g_{n-1}, H_{n-1} g_{n-1} \rangle} \end{aligned}$$

Therefore, since H_{n-1} is positive-definite, we must have

$$\langle g_m, H_{n-1}g_n\rangle = 0 \quad \text{for} \quad n < m$$

But then, from the definitions of H_n and B_n, we have for $n < m$ that

$$H_n g_m = H_{n-1} g_m = \ldots = H_0 g_m$$

and hence g_m is an eigenvector with eigenvalue 1 for $H_0^{-1}H_n$ if $m > n$. Therefore, we find that

$$p_n = \frac{[\langle g_n, H_0 g_n\rangle I + H_0 g_n g_{n-1}^*]p_{n-1}}{\langle g_n, H_0 g_n\rangle + \langle g_{n-1}, p_{n-1}\rangle}$$

For the conjugate-gradient method, $p_0' = Kg_0 = p_0$, and therefore $x_1' = x_1$. If $p_i = \lambda_i p_i'$ and $x_{i+1}' = x_{i+1}$ for $i = 0, 1, \ldots, n$, as is true for $n = 0$, then

$$p_{n+1} = \frac{[\langle g_{n+1}, H_0 g_{n+1}\rangle I + H_0 g_{n+1} g_n^*]\lambda_n p_n'}{\langle g_{n+1}, H_0 g_{n+1}\rangle + \langle g_n, p_n\rangle}$$

while

$$\begin{aligned}
p_{n+1}' &= H_0 g_{n+1} + \frac{\langle g_{n+1}, H_0 g_{n+1}\rangle}{\langle g_n, H_0 g_n\rangle} p_n' \\
&= \frac{\langle g_n, H_0 g_n\rangle H_0 g_{n+1} + \langle g_{n+1}, H_0 g_{n+1}\rangle p_n'}{\langle g_n, H_0 g_n\rangle} \\
&= \frac{\langle g_{n+1}, H_0 g_{n+1}\rangle + \langle g_n, p_n\rangle}{\lambda_n \langle g_n, H_0 g_n\rangle} \\
&\quad \times \left\{ \frac{\langle g_{n+1}, H_0 g_{n+1}\rangle \lambda_n p_n' + H_0 g_{n+1}\langle g_n, \lambda_n p_n'\rangle}{\langle g_{n+1}, H_0 g_{n+1}\rangle} \right. \\
&\qquad \left. \frac{+ \lambda_n \langle g_n, H_0 g_n\rangle H_0 g_{n+1} - H_0 g_{n+1}\langle g_n, \lambda_n p_n'\rangle}{+ \langle g_n, p_n\rangle} \right\} \\
&\equiv \frac{1}{\lambda_{n+1}} p_{n+1}
\end{aligned}$$

(where λ_{n+1} is defined by the equality) since

$$\langle g_n, H_0 g_n\rangle = \langle g_n, p_n'\rangle - b_{n-1}\langle g_n, p_{n-1}\rangle = \langle g_n, p_n'\rangle$$

Thus $p_{n+1} = \lambda_{n+1} p_{n+1}'$ and hence $x_{n+2} = x_{n+2}'$. Q.E.D.

Thus, because of Theorems 7.4.2 and 7.4.3 and the results of Chapter 5 and Section 6.4, we know that the methods of Equation 7.4.1 (and in particular, Davidon's method) yield global convergence when implemented with exact minimization along $x_n + tp_n$ and applied to uniformly convex quadratic functionals—for another proof see [Horwitz–Sarachik (1968)]—and we have estimates on the local-convergence rate. Similarly, for x_0 near x^*, for uniformly convex nonquadratic functionals, their relationship to conjugate-gradient methods gives a local-convergence result for these methods as well. Only very recently [Powell (1970)] has a global-convergence result for the Davidon method been found; we give this result and outline the proof.

THEOREM 7.4.4 [Powell (1970)]. Let f be twice continuously differentiable, let $f''_x \geq aI$, $a > 0$, and let the Davidon algorithm be used with exact minimization along $x_n + tp_n$, starting with an arbitrary x_0 and positive-definite symmetric H_0. Then the sequence $\{x_n\}$ converges to the unique solution $\hat{x}$.

Proof: Since H_n is positive-definite for each n [Fletcher–Powell (1963)], we can define $\mathbf{\Gamma}_n \equiv H_n^{-1}$. Writing

$$\gamma_n \equiv x_{n+1} - x_n$$

one can check that

$$\mathbf{\Gamma}_{n+1} = \left(I - \frac{\delta_n \gamma_n^*}{\langle \gamma_n, \delta_n \rangle}\right)\mathbf{\Gamma}_n\left(I - \frac{\gamma_n \delta_n^*}{\langle \gamma_n, \delta_n \rangle}\right) + \frac{\delta_n \delta_n^*}{\langle \gamma_n, \delta_n \rangle}$$

Writing $\operatorname{tr}(A)$ for the *trace* of a matrix A—that is, $\operatorname{tr}(A) = \sum_{i=1}^{l} A_{ii}$, which equals the sum of the eigenvalues for symmetric A—we have

$$\operatorname{tr}(\mathbf{\Gamma}_{n+1}) = \operatorname{tr}(\mathbf{\Gamma}_n) - 2\frac{\langle \gamma_n, \mathbf{\Gamma}_n \delta_n \rangle}{\langle \gamma_n, \delta_n \rangle} + \frac{\langle \gamma_n, \mathbf{\Gamma}_n \gamma_n \rangle \langle \delta_n, \delta_n \rangle}{\langle \gamma_n, \delta_n \rangle^2} + \frac{\langle \delta_n, \delta_n \rangle}{\langle \gamma_n, \delta_n \rangle}$$

Using the fact that

$$\langle \nabla f(x_{n+1}), \gamma_n \rangle = t_n \langle \nabla f(x_{n+1}), p_n \rangle = 0$$

and

$$\frac{1}{\langle \nabla f(x_{n+1}), H_{n+1} \nabla f(x_{n+1}) \rangle} = \frac{1}{\langle \nabla f(x_{n+1}), H_n \nabla f(x_{n+1}) \rangle} + \frac{1}{\langle \nabla f(x_n), H_n \nabla f(x_n) \rangle}$$

we conclude that

$$\operatorname{tr}(\mathbf{\Gamma}_{n+1}) = \operatorname{tr}(\mathbf{\Gamma}_n) + \frac{\|\nabla f(x_{n+1})\|^2}{\langle \nabla f(x_{n+1}), H_{n+1} \nabla f(x_{n+1}) \rangle} - \frac{\|\nabla f(x_n)\|^2}{\langle \nabla f(x_n), H_n \nabla f(x_n) \rangle} - \frac{\|\nabla f(x_{n+1})\|^2}{\langle \nabla f(x_{n+1}), H_n \nabla f(x_{n+1}) \rangle} + \frac{\|\delta_n\|^2}{\langle \gamma_n, \delta_n \rangle}$$

Solving this recursion thus yields

$$\operatorname{tr}(\Gamma_{n+1}) = \operatorname{tr}(\Gamma_0) + \frac{\|\nabla f(x_{n+1})\|^2}{\langle \nabla f(x_{n+1}), H_{n+1}\nabla f(x_{n+1})\rangle} - \frac{\|\nabla f(x_0)\|^2}{\langle \nabla f(x_0), H_0 \nabla f(x_0)\rangle}$$
$$- \sum_{i=0}^{n} \frac{\|\nabla f(x_{i+1})\|^2}{\langle \nabla f(x_{i+1}), H_i \nabla f(x_{i+1})\rangle} + \sum_{i=0}^{n} \frac{\|\delta_i\|^2}{\langle \gamma_i, \delta_i\rangle}$$

A simple use of the mean-value theorem implies that

$$\frac{\|\delta_i\|^2}{\langle \gamma_i, \delta_i\rangle} \leq B_0$$

for some constant B_0. Since

$$\frac{\|z\|^2}{\langle z, H_{n+1} z\rangle}$$

is less than or equal to the largest eigenvalue of $H_{n+1}^{-1} = \Gamma_{n+1}$, which is less than $\operatorname{tr}(\Gamma_{n+1})$, and since

$$\operatorname{tr}(\Gamma_0) \quad \text{and} \quad \frac{\|\nabla f(x_0)\|^2}{\langle \nabla f(x_0), H_0 \nabla f(x_0)\rangle}$$

are constants, we find that a B exists such that

$$\sum_{i=0}^{n} \frac{\|\nabla f(x_{i+1})\|^2}{\langle \nabla f(x_{i+1}), H_i \nabla f(x_{i+1})\rangle} < Bn \tag{7.4.2}$$

Arguing similarly for H_n instead of Γ_n and using $\operatorname{tr}(H_{n+1}) \geq 0$, we can also conclude that there is a constant M such that

$$\sum_{i=0}^{n} \frac{\|H_i \delta_i\|^2}{\langle \delta_i, H_i \delta_i\rangle} < Mn$$

Now, using

$$\langle \delta_i, H_i \delta_i\rangle^2 \leq \|H_i \delta_i\|^2 \|\delta_i\|^2$$

and

$$\langle \delta_i, H_i \delta_i\rangle > \langle \nabla f(x_{i+1}), H_i \nabla f(x_{i+1})\rangle$$

we finally conclude that

$$\sum_{i=0}^{n} \frac{\langle \nabla f(x_{i+1}), H_i \nabla f(x_{i+1})\rangle}{\|\delta_i\|^2} < Mn \tag{7.4.3}$$

Thus, from Equation 7.4.2, for at least two-thirds of the integers i in $0 \leq i \leq n$, we must have

$$\frac{\|\nabla f(x_{i+1})\|^2}{\langle \nabla f(x_{i+1}), H_i \nabla f(x_{i+1})\rangle} \leq 3B$$

From Equation 7.4.3, at least two-thirds of the terms must also satisfy

$$\frac{\langle \nabla f(x_{i+1}), H_i \nabla f(x_{i+1})\rangle}{\|\delta_i\|^2} \leq 3M$$

Therefore, for at least one-third of the integers i in $0 \leq i \leq n$, we must have both of the above inequalities—that is,

$$\begin{aligned}\|\nabla f(x_{i+1})\|^2 &\leq 3B\langle \nabla f(x_{i+1}), H_i \nabla f(x_{i+1})\rangle \\ &\leq 9BM\|\delta_i\|^2 = 9BM\frac{\|\delta_i\|^2}{\|\gamma_i\|^2}\|\gamma_i\|^2\end{aligned}$$

Since, by the mean-value theorem,

$$\frac{\|\delta_i\|^2}{\|\gamma_i\|^2}$$

is bounded and $f(x_i) - f(x_{i+1}) \geq \frac{1}{2}a\|\gamma_i\|^2$ must converge to zero, we deduce the existence of an infinite subsequence of x_{i_j} such that $\|\nabla f(x_{i_j})\| \longrightarrow 0$. It follows from the uniform convexity of f that $x_{i_j} \longrightarrow \hat{x}$; since $f(x_n) \leq f(x_{i_j})$ if $n \geq i_j$, we then deduce that the entire sequence $\{x_n\}$ converges to $\hat{x}$. Q.E.D.

So far as the local-convergence rate is concerned, one can also prove the following theorem.

THEOREM 7.4.5 [Powell (1970)]. Under the hypotheses of Theorem 7.4.4, there exists a scalar $\alpha \in (0, 1)$ such that

$$f(x_n) - f(\hat{x}) \leq \alpha^n[f(x_0) - f(\hat{x})]$$

If in addition f''_x satisfies a uniform Lipschitz condition in $\{x; f(x) \leq f(x_0)\}$, then there exist $a > 0$, $A < \infty$ such that $aI \leq H_n \leq AI$, and

$$\frac{\|x_{n+1} - \hat{x}\|}{\|x_n - \hat{x}\|}$$

converges to zero; $\{H_n\}$ need not converge to $[f''_{\hat{x}}]^{-1}$.

The local-convergence rate has also been studied in McCormick (1969), where somewhat weaker results than those above are obtained; an analysis of a *restart Davidon method*—in which, after every l steps, H_n is reset to H_0—is also presented there. A computer program for the Davidon method can be found in Wells (1965).

EXERCISE. Prove global convergence for the restart Davidon method.

A method similar to the Davidon method attempts to compute the inverse of f''_x more directly. If one has proceeded through l successive directions $p_1, \ldots, p_l$ and defines the matrix P with columns $p_1, \ldots, p_l$, then for a quadratic functional

$$f(x) = \langle h - x, M(h - x)\rangle$$

we have $P^*MP = D$ so that $M^{-1} = PD^{-1}P^*$. Thus in general a reasonable approximation for $(f''_x)^{-1}$, given l directions $p_1, \ldots, p_l$, would be

$$H = PD^{-1}P^*$$

Of course, one cannot actually *compute* D without first computing the Hessian matrix, which we seek to avoid. However, we note for quadratics that $D = P^*MP = P^*GT^{-1}$ where G is a matrix whose ith column is $\delta_i = \nabla f(x_{i+1}) - \nabla f(x_i)$ and T is a diagonal matrix having as its diagonal $(t_1, \ldots, t_l)$. Defining G thus in terms of differences of gradients for arbitrary f, we can then take as a good initial matrix the matrix

$$H_0 = PTG^{-1}$$

It has been suggested [Ritter (1969)] that one use essentially this choice of the H-matrix at every step. The point of this is that even for nonquadratic functionals, setting $H_n = P_nT_nG_n^{-1}$ where

$$P_n = (p_{n-l}, p_{n-l+1}, \ldots, p_{n-1}),$$
$$G_n = (\delta_{n-l}, \ldots, \delta_{n-1})$$

and

$$T_n = \operatorname{diag}(t_{n-l}, \ldots, t_{n-1})$$

guarantees that

$$H_n\delta_i = x_{i+1} - x_i \quad \text{for} \quad n - l \le i \le n - 1$$

while this relationship holds for the Davidon method only for quadratics. To get global convergence for this method, it has been suggested [Ritter (1969)] that p_n be chosen as $-H_n\nabla f(x_n)$ if

$$\langle -\nabla f(x_n), H_n\nabla f(x_n)\rangle \ge \epsilon \|\nabla f(x_n)\|^2 \quad \text{for a fixed } \epsilon > 0$$

and as, say, $-\nabla f(x_n)$ or some other globally convergent choice otherwise. This clearly gives global convergence for sufficiently convex problems, assuming that a reasonable t_n is chosen.

EXERCISE. Prove the global convergence asserted in the preceding paragraph.

If the choice of H_n is modified so that it is computed from a "sufficiently linearly independent" set of l preceding directions—say, by requiring $\det(P_n) \geq \epsilon$—the local-convergence properties can be analyzed. In particular [Ritter (1969)], x_n converges superlinearly to $\hat{x}$ and H_n converges to $f''^{-1}_{\hat{x}}$. Computationally, one would not want to be forced to compute G_n^{-1} every time; fortunately, it is simple to compute G_{n+1}^{-1} recursively from G_n^{-1} using the relationship between G_{n+1} and G_n. This appears to be an excellent method in theory, but we are not aware of any significant computational results.

EXAMPLE [Kowalik–Osborne (1969)]. Consider using Davidon's variable-metric method with $H_0 = I$ to minimize

$$f(x, y) = 100(y - x^2)^2 + (1 - x)^2$$

whose unique minimum is at $x = y = 1$. Letting $x_0 = -1.2$, $y_0 = 1.0$, 50 functional calls of f were required to find $(x, y) = (0.6885, 0.4580)$, $f = 0.122$, and 80 calls yielded $(x, y) = (0.992, 0.9984)$, $f = 6.27 \times 10^{-7}$.

Another particular variable-metric method [Broyden (1965)] is obtained by setting

$$q_n = z_n = \frac{H_n p_n}{\langle H_n p_n, \delta_n \rangle}$$

This method is not exact, but if applied to $\langle h - x, M(h - x)\rangle$, one finds

$$\| H_{n+1}^{-1} - M \| \leq \| H_n^{-1} - M \|$$

in the spectral norm. Experience with the method is somewhat limited, but it does appear to be of value and deserving of further study. Applying one implementation of the method [Broyden (1965)] to the example immediately above, with $x_0 = -1.2, y_0 = 1.0$, gave $f = 10^{-19}$ with 59 function evaluations.

A method for which some information is available as to the admissibility of the direction sequence in some special cases is one due to P. Wolfe and presented by Goldfarb [Goldfarb (1969b)]. In this case, one takes

$$z_n = q_n = \frac{H_n\delta_n - t_n p_n}{\langle H_n\delta_n - t_n p_n, \delta_n \rangle}$$

$$H_{n+1} = H_n + B_n$$

if no division by zero is required, and $H_{n+1} = H_n$ otherwise. For convenience we write $A > B\,(A \geq B)$ to mean that $A - B$ is positive-definite (semi-definite). By straightforward arguments one can show that for this method with

$$f(x) = \langle h - x, M(h - x)\rangle$$

if $H_0 \geq M^{-1}$ $(H_0 \leq M^{-1})$, then

$$H_0 \geq H_n \geq M^{-1}(H_0 \leq H_n \leq M^{-1}) \quad \text{for all} \quad n$$

For a general nonquadratic functional f, if $0 < aI \leq f''_x \leq AI$, if $f''^{-1}_{x_0} \geq H_0 \geq kI$ for some $k \leq 1/A$, and if $f''_{x+p} \leq f''_x$ whenever $f(x + p) \leq f(x)$, then $kI \leq H_n \leq (1/a)I$ for all n. This then states that the direction sequence, under these hypotheses, is admissible and hence that the iterative method will yield a criticizing sequence, in this case converging to $\hat{x}$. A similar theorem is valid for $H_0 \geq f''^{-1}_{x_0}$. To the author's knowledge this is the *only* result on the admissibility of a variable-metric direction sequence. The statement of the result with $H_0 \leq f''^{-1}_{x_0}$ is in fact useful computationally, since this is satisfied with $H_0 = \lambda^{-1}I$ where $\lambda \geq \| f''_{x_0} \|$; thus if f''_x is nonincreasing along paths of nonincreasing functional values, such a computable choice of H_0 will assure convergence.

Some other two-parameter–type methods have been considered briefly in the literature [Broyden (1967), Vercoustre (1969)] but have not proved particularly successful; for example only, we mention the possibilities

$$q_n = z_n = \frac{p_n}{\langle p_n, \delta_n\rangle} \quad \text{and} \quad q_n = z_n = \frac{H_n\delta_n}{\langle H_n\delta_n, \delta_n\rangle}$$

7.5. CONSTRAINED PROBLEMS

Because of the experimental evidence and, in some few cases, theoretical analyses showing that variable-metric and in particular the conjugate-gradient and Davidon methods are extremely powerful tools for unconstrained minimization, a great deal of interest has arisen lately in variable-metric methods for constrained minimization; conjugate-direction methods have been applied in various ways [Barnes (1965,) Goldfarb (1966, 1969a), Goldfarb–Lapidus (1968), Zoutendijk (1960)], perhaps the most promising of which is an analogue of the Davidon–Fletcher–Powell method. In the unconstrained case for a quadratic f, this method merely uses the steepest-descent direction defined by a metric approximating f''. An analogue for constrained minimization is to use the projected-gradient method where the projection is with respect to some different metric. This of course can be described in the variable-metric setting with $p_n = -H_n \nabla f(x_n)$ where H_n is a projection matrix for a varying metric. Such an algorithm can be made to be exact for quadratics with linear constraints. This Davidon-type method for constrained problems has not been thoroughly analyzed with respect to its global-convergence properties. Evidence indicates that this will be a very powerful method, but much of the theoretical analysis is lacking and should be provided. The theory that is known can be found in Goldfarb–Lapidus (1968) and in part in Section 4.10, particularly Theorem 4.10.2 of this volume.

8 OPERATOR-EQUATION METHODS

8.1. INTRODUCTION

The methods discussed in the preceding chapters have all been of gradient type—that is, such that $\langle x_{n+1} - x_n, \nabla f(x_n)\rangle < 0$; these are the types of methods which arise most naturally for unconstrained minimization problems. Another large class of useful methods, however, is comprised of those which in essence ignore the minimization aspect of the problem and treat directly the operator equation

$$\nabla f(x) = 0$$

A solution to this equation need not of course minimize f unless f is further restricted—for example, by some convexity hypotheses. We cannot hope to describe in any detail the tremendous variety of methods available for the solution of such equations; entire books, and large sections of others, are devoted to these methods [Collatz (1966), Ostrowski (1966b), Rall (1969), Traub (1964)]. We shall, therefore, be very brief and merely mention the various methods which appear to be useful and the kinds of problems to which their application is understood; only the barest outline is presented here.

We are primarily interested in methods for which global-convergence results are known; that is, we wish to locate a solution x^* iteratively starting with any x_0 whatsoever, or at least with restricted x_0 that can easily be found. Although local-convergence results have long been known for many iterative methods, global results are much harder to obtain and are commensurably less common. Global results usually are obtained either by variational analysis or by an analysis using some type of abstract, generalized monotonicity viewpoint. From the variational viewpoint, some methods are shown to be of gradient type and then the general analysis of the preceding chapters can be applied to analyze convergence; we have already seen this, for example, for Newton's method. We shall look further into Newton's method from both viewpoints in this chapter, examine briefly the iterative methods generalizing well-known methods for linear equations—such as Jacobi, Gauss–Seidel, successive-relaxation, and alternating-direction iterations—and conclude with a study of methods based on minimizing $||\nabla f(x)||^2$—that is, on least-squares methods. Let us then turn to the question of iterative solution of nonlinear operator equations

$$J(x) = 0$$

8.2. NEWTON AND NEWTON-LIKE METHODS

The original *Newton method* for nonlinear equations takes the form

$$x_{n+1} = x_n - J_n'^{-1}J(x_n)$$

where as usual we write J_n' for the Frechet derivative at x_n; that is, $J_n' = J_{x_n}'$. We have mentioned briefly the use of Newton's method when $J(x) = \nabla f(x)$ for a function f to be minimized, in which case we take the direction $p_n = -J_n'^{-1}J(x_n)$ and let $x_{n+1} = x_n + t_n p_n$ where t_n is determined by one of the general methods of Chapter 4; we know that choosing t_n to minimize $f(x_n + tp_n)$, for example, yields quadratic convergence. We can show in addition that, roughly speaking, one can always eventually choose t_n equal to 1, in which case we have the pure Newton method which also converges quadratically [Kantorovich–Akilov (1964), Rall (1969)].

THEOREM 8.2.1. Suppose $J(x) \equiv \nabla f(x)$ is such that $0 < aI \leq J_x' \leq AI$ for all x in $W(x_0)$, the closed convex hull of $\{x; f(x) \leq f(x_0)\}$, and suppose that J_x' is uniformly continuous in x in $W(x_0)$. Let

$$x_{n+1} = x_n - t_n J_n'^{-1}J(x_n)$$

where, for a fixed α in $(0, 1)$, t_n is the first of the numbers $t = \alpha^0, \alpha^1, \alpha^2, \ldots,$ satisfying

$$f(x_n) - f(x_n + tp_n) \geq qt_n \langle J(x_n), J_n'^{-1}J(x_n)\rangle$$

for a fixed q in $(0, \frac{1}{2})$. Then $x_n \longrightarrow \hat{x}$, the unique point minimizing f. For sufficiently large n, we may take $t_n = 1$ and hence the convergence is quadratic.

Proof: Since $W(x_0)$ is bounded and $||J_x'|| \leq A$, ∇f is uniformly continuous in $W(x_0)$. We have

$$\frac{a}{A}||\nabla f(x_n)|| \leq \left\langle -\nabla f(x_n), \frac{p_n}{||p_n||}\right\rangle \leq A||p_n||$$

and hence the assumptions of Corollary 4.7.2 are satisfied with $d(t) \equiv \delta t$, $d_1(t) \equiv (1/A)t$. We thus conclude that

$$\nabla f(x_n) \longrightarrow 0$$

Then by Theorems 4.2.1, 1.5.1, and 1.6.1 we conclude that $x_n \longrightarrow \hat{x}$. To show that we may take $t_n = 1$ for large n, we write

$$\begin{aligned}
\frac{f(x_n) - f(x_n + p_n)}{\langle J(x_n), J_n'^{-1}J(x_n)\rangle} &= \frac{f(x_n) - [f(x_n) + \langle \nabla f(x_n), p_n\rangle + \frac{1}{2}\langle J'_n p_n, p_n\rangle]}{\langle J(x_n), J_n'^{-1}J(x_n)\rangle} \\
&\quad + \frac{\frac{1}{2}\langle J'_n p_n, p_n\rangle - \frac{1}{2}\langle J'_{x_n+\lambda_n p_n} p_n, p_n\rangle}{\langle J(x_n), J_n'^{-1}J(x_n)\rangle} \quad \text{for some} \quad \lambda_n \in (0, 1) \\
&= \frac{\frac{1}{2}\langle J(x_n), J_n'^{-1}J(x_n)\rangle + o(\|J(x_n)\|^2)}{\langle J(x_n), J_n'^{-1}J(x_n)\rangle} \\
&= \tfrac{1}{2} + o(1)
\end{aligned}$$

Since $q \in (0, \frac{1}{2})$, for large n we may take $t_n = 1$, thus yielding the well-known quadratic convergence. Q.E.D.

The above theorem shows how variational analysis can be used to prove a global-convergence theorem for (an asymptotically null modification of) Newton's method. We next state a similar result proved by monotonicity methods [Baluev (1952), Ortega–Rheinboldt (1967a)]; for simplicity we restrict ourselves to $\mathbb{R}^l$. We remark that if x and y are two elements of a finite-dimensional space, $x \geq y$ is defined to mean that the components of $x - y$ are nonnegative.

THEOREM 8.2.2. Let $J: \mathbb{R}^l \longrightarrow \mathbb{R}^l$ be continuously differentiable and satisfy

$$J[\lambda x + (1 - \lambda)y] \leq \lambda J(x) + (1 - \lambda)J(y) \quad \text{for} \quad x, y \quad \text{in} \quad \mathbb{R}^l, \qquad \lambda \in [0, 1]$$

Let $J_x'^{-1}$ exist, and let $J_x'^{-1} \geq 0$ for all x in $\mathbb{R}^l$. If $J(x) = 0$ has a solution $\hat{x}$ in $\mathbb{R}^l$, then it is unique and for any x_0 in $\mathbb{R}^l$ the sequence $x_{n+1} = x_n - J_n'^{-1}J(x_n)$ converges to $\hat{x}$ and satisfies $x_n \geq x_{n+1} \geq \hat{x}$ for $n \geq 1$.

Proof: The "convexity" assumption on J merely says that J is convex in each of its components; thus from Proposition 1.5.1 we have

$$J(x) - J(y) \geq J'_y(x - y)$$

If $J(x) = J(y) = 0$, then one has

$$0 \geq J'_y(x - y)$$

and since $J_y'^{-1} \geq 0$, we have

$$x - y \leq 0$$

Similarly, $y - x \leq 0$ and hence $x = y$—that is, $\hat{x}$ is unique. Now, for any x_0,

$$\begin{aligned} x_1 - \hat{x} &= x_0 - J_0'^{-1}J(x_0) - [\hat{x} - J_0'^{-1}J(\hat{x})] \\ &= x_0 - \hat{x} + J_0'^{-1}[J(\hat{x}) - J(x_0)] \\ &\geq x_0 - \hat{x} + J_0'^{-1}[J_0'(\hat{x} - x_0)] \\ &\geq 0 \end{aligned}$$

Therefore, since x_0 is arbitrary, $x_n \geq \hat{x}$ for $n \geq 1$. But then

$$-J(x_{n-1}) = J'_{n-1}(x_n - x_{n-1}) \leq J(x_n) - J(x_{n-1})$$

and hence

$$J(x_n) \geq 0 \quad \text{for} \quad n \geq 1$$

yielding

$$x_{n+1} - x_n = -J_n'^{-1}J(x_n) \leq 0$$

Thus we have

$$x_n \geq x_{n+1} \geq \hat{x} \quad \text{for} \quad n \geq 1$$

Therefore, there exists at least one convergent subsequence; for any such subsequence $\{x_{n_i}\}$, with $x_{n_i} \longrightarrow x'$, since

$$x_{n_{i+1}} \leq x_{n_i+1} \leq x_{n_i}$$

it is easy to show that $x_{n_i+1} \longrightarrow x'$ also and hence $J(x') = 0$. Therefore, all limit points of $\{x_n\}$ must equal $\hat{x}$ and hence $x_n \longrightarrow \hat{x}$. Q.E.D.

Since Newton's method is often cumbersome because of the need to compute the derivatives in J'_n and to solve a linear system, one often desires to use modifications of the method, such as

$$x_{n+1} = x_n - H_n J(x_n)$$

where H_n is some "approximation" to $J_n'^{-1}$; thus we are back to the *quasi-Newton methods* (called variable-metric methods when considered from the variational viewpoint). If we ask only local-convergence theorems, these methods have been thoroughly studied in general. A typical such theorem for $H_n = B(x_n)^{-1}$ is the following [Dennis (1969), Rheinboldt (1967)].

PROPOSITION 8.2.1. Let $J\colon X \longrightarrow Y$ for Banach spaces X, Y, $\|J'_x - J'_y\| \leq K\|x - y\|$ for $x, y \in D_0$, a convex subset of X. Let $B\colon D_0 \longrightarrow L[X, Y]$, and for a given $x_0 \in D_0$, let

$$\|B(x) - B(x_0)\| \leq \eta\|x - x_0\|, \quad \|J'_x - B(x)\| \leq \delta_0 + \delta_1\|x - x_0\|$$

for x in D_0. Let

$$\|B^{-1}(x_0)\| \leq \beta, \|B^{-1}(x_0)J(x_0)\| \leq \alpha,$$

and assume that $\beta\delta_0 < 1$ and

$$h \equiv \frac{\sigma\beta K\alpha}{(1-\beta\delta_0)^2} \leq \frac{1}{2}$$

where

$$\sigma = \max\left(1, \frac{\eta + \delta_1}{K}\right)$$

Set

$$r_1 = \frac{1-\sqrt{1-2h}}{h} \times \frac{\alpha}{1-\beta\delta_0}, \qquad r_2 = \frac{1+\sqrt{1-(2h/\sigma)}}{h} \times \frac{\sigma\alpha}{1-\beta\delta_0}$$

If $S(x_0, r_1) \subset D_0$, then the sequence

$$x_{n+1} = x_n - B(x_n)^{-1}J(x_n)$$

exists, lies in $S(x_0, r_1)$, and converges to a solution $\hat{x}$ of $J(x) = 0$ unique inside $S(x_0, r_2) \cap D_0$.

We are looking, however, for theorems about global convergence, not local; to our knowledge such theorems other than contraction-mapping theorems are not known (except for the variational results of earlier chapters) for the general quasi-Newton method. By making much more specialized assumptions on the precise form of the method, some global results have been obtained; this is the subject, in effect, of the next sections.

We wish to point out that some of the quasi-Newton methods are implemented with a variational flavor: given x_n and a direction p_n, one sets $x_{n+1} = x_n + t_n p_n$ where t_n is chosen to minimize or reduce some local-error measure $E_n(x)$; for the pure variational methods we have of course $E_n(x) \equiv f(x)$ for all n. A particularly popular choice in general is $E_n(x) = \|J(x)\|^2$. The general form of the conjugate-gradient method for nonlinear operators, mentioned in Section 7.3, treats the error measure $E_n(x) = \langle J(x), H_{x_n}J(x)\rangle$ for some operator H_{x_n}. The direction is

$$p_n = -K_{x_n}J(x_n) + b_{n-1}p_{n-1}$$

where K_{x_n} is another operator and b_{n-1} is chosen to make p_n and p_{n-1} $(J_n'^*H_{x_n}J_n')$-conjugate [Daniel (1965; 1967a, b)]. This exhibits very rapid convergence near the solution, but also reveals the difficult problem of such methods—namely, making a good choice for the operator H_{x_n}. In Daniel

(1967a), an example of two equations in two unknowns is handled in this way with two different, apparently reasonable error measures, with very different results; one method gives $||J(x_{40})|| \leq 0.007\,||J(x_0)||$, while the other gives $||J(x_{27})|| \leq 10^{-6}\,||J(x_0)||$. Thus this type of variable error-measure method, while seeming to grow in popularity, is but poorly understood in general.

8.3. GENERALIZED LINEAR ITERATIONS

We next consider some iterative methods which are natural extensions of well-known methods for linear equations. Let vectors x in $\mathbb{R}^l$ have components $\xi_1, \ldots, \xi_l$, and consider the linear system

$$Ax = b$$

The following iterative relaxation methods are well known [Douglas–Kellogg–Varga (1963), Varga (1962)].

Jacobi method: Given x_n, compute x_{n+1} via

$$a_{kk}\xi_{n+1,k} = b_k - \sum_{\substack{j=1 \\ j \neq k}}^{l} a_{kj}\xi_{n,j}, \qquad k = 1, 2, \ldots, l$$

There is also a *block* version in which several $\xi_{n+1,k}$ are found simultaneously.

Successive overrelaxation (SOR) method: Given x_n, compute x_{n+1} by letting

$$a_{kk}\xi'_k = b_k - \sum_{j=1}^{k-1} a_{kj}\xi_{n+1,j} - \sum_{j=k+1}^{l} a_{kj}\xi_{n,j}$$

$$\xi_{n+1,k} = (1 - w)\xi_{n,k} + w\xi'_k, \qquad k = 1, 2, \ldots, l$$

for some *relaxation factor* w. There is also a block version in which several $\xi_{n+1,k}$ are found simultaneously.

Gauss–Seidel method: This is just SOR with the relaxation parameter $w \equiv 1$.

Alternating direction implicit (ADI) method: Suppose $A = H + V$. Given x_n and some parameters $s_n > 0$ and $r_n > 0$, compute x_{n+1} via

$$(s_nI + H)x_{n+(1/2)} = (s_nI - V)x_n + b$$

$$(r_nI + V)x_{n+1} = (r_nI - H)x_{n+(1/2)} + b$$

There are many ways in which these methods can be generalized to nonlinear problems; we mention three common approaches [Ortega–Rheinboldt

(1968)]. Let $\backsim$ be the *name* of one of the methods above having the form $x_{n+1} = F(x_n; A; b)$.

1. *Nonlinear $\backsim$ method.* In each of the first three of the above methods, one or more components of x_{n+1} were computed successively; for the nonlinear version, precisely the same is done. Thus to solve $G(x) = 0$ where $G: \mathbb{R}^l \to \mathbb{R}^l$, $G = (g_1, \ldots, g_l)^*$, the *nonlinear Jacobi method*, for example, yields x_{n+1} from x_n via

$$g_k(\xi_{n,1}, \ldots, \xi_{n,k-1}, \xi_{n+1,k}, \xi_{n,k+1}, \ldots, \xi_{n,l}) = 0, \qquad k = 1, \ldots, l$$

For *nonlinear ADI*, we assume $G(x) = H(x) + V(x)$, and then the method looks just like the linear one.

EXERCISE. Write out the nonlinear Jacobi method for solving $\xi_1 + \xi_2^2 = 0$, $e^{\xi_1} - \xi_2 = 4$.

2. *$\backsim$-m-step-Newton method.* In each nonlinear $\backsim$ method, it was necessary to solve at least l nonlinear equations to get from x_n to x_{n+1}. Since one cannot do this exactly, one might try to compute x_{n+1} by using m steps of Newton's method on the required nonlinear problem [Lieberstein (1959)]. Thus the *Jacobi–1-step–Newton method* is

$$\xi_{n+1,k} = \xi_{n,k} - \frac{g(x_n)}{\dfrac{\partial g}{\partial \xi_k}(x_n)}, \qquad k = 1, 2, \ldots, l$$

EXERCISE. Write out the Jacobi–1-step–Newton method for solving $\xi_1 + \xi_2^2 = 0$, $e^{\xi_1} - \xi_2 = 4$.

3. *Newton–m-step–$\backsim$ method.* A third possibility is to consider solving $G(x) = 0$ by Newton's method:

$$x_{n+1} = x_n - G_n'^{-1}G(x_n)$$

Since this involves solving a linear system at each step, we might consider using m steps of a method $F(x_n; A; b)$ for solving the linear system, thus getting the method

$$x_{n+1} = F^m[x_n; G_n'; G_n'x_n - G(x_n)]$$

Thus the *Newton–1-step–Gauss–Seidel method* is

$$\frac{\partial g_k}{\partial \xi_k}(x_n)\xi_{n+1,k} = \sum_{j=1}^{k-1} \frac{\partial g_k}{\partial \xi_j}(x_n)(\xi_{n,j} - \xi_{n+1,j}) - g_k(x_n), \quad k = 1, 2, \ldots, l$$

EXERCISE. Write out the Newton–1-step–Jacobi method for solving $\xi_1 + \xi_2^2 = 0$, $e^{\xi_1} - \xi_2 = 4$.

The local convergence of such methods can be analyzed in the usual way by using the fact that, near solutions, the nonlinearities contribute only second-order effects; thus many of the results for linear equations carry over without change to the local convergence for nonlinear problems [Ortega–Rockoff (1966), Ortega–Rheinboldt (1968)]. The global-convergence properties are far more difficult, however. With the exception of the ADI-type methods, some useful results can be obtained from the variational viewpoint, especially for the pure nonlinear methods. This is because the step from x_n to x_{n+1} can be considered as the result of l steps through the coordinate directions, which directions will be of gradient type for certain problems, thus allowing the application of any of the various step-size algorithms. The following proposition is typical of the type of result that has been obtained from the variational viewpoint.

PROPOSITION 8.3.1 [Schechter (1962, 1968)]. Let $f(x)$ be twice continuously differentiable in $\mathbb{R}^l$ with $0 < aI \leq f''_x$ for all x. Then $\nabla f(x) = 0$ has a unique solution and for any intitial x_0 the sequence generated by the nonlinear Gauss–Seidel method converges to that solution.

This result has been refined and extended in many ways—e.g., to Jacobi, SOR, and block methods; some similar results are known for the n–1-step–Newton methods also [Elkin (1968), Ortega–Rheinboldt (1968), Schechter (1968)]. For nearly all of the other generalized linear iterations, monotonicity methods have been most successful for treating global-convergence questions. We mention briefly some of these results. We recall that an *M-matrix* $A = ((a_{ij}))$ is one such that $A^{-1} \geq 0$ and $a_{ij} \leq 0$ if $i \neq j$, and that $\phi: \mathbb{R}^l \longrightarrow \mathbb{R}^l$ is *isotone* if $x \leq y$ implies $\phi(x) \leq \phi(y)$; is *diagonal* if, for $1 \leq i \leq l$, the ith component of $\phi(x)$ depends only on the ith component of ϕ; and is *convex* if

$$\phi[\lambda x + (1 - \lambda)y] \leq \lambda\phi(x) + (1 - \lambda)\phi(y)$$

for all x, y in $\mathbb{R}^l$. Such operators commonly arise in the numerical solution of boundary-value problems for mildly nonlinear differential equations [Bers (1953), Greenspan–Parter (1965), Schechter (1962)].

The following two results are typical of what is known for the full nonlinear methods.

PROPOSITION 8.3.2 [Bers (1953), Ortega–Rheiboldt (1967a, 1968)]. Let A be an M-matrix, $\phi: \mathbb{R}^l \longrightarrow \mathbb{R}^l$ a continuous, diagonal isotone mapping, and set

$$G(x) = Ax + \phi(x)$$

Then for any x_0 in $\mathbb{R}^l$, the nonlinear Jacobi and nonlinear SOR [for

$0 < w \leq 1$ (and hence, for $w = 1$, the nonlinear Gauss–Seidel)] methods all yield sequences x_n converging to the unique solution $\hat{x}$ of $G(x) = 0$.

PROPOSITION 8.3.3 [Caspar (1968), Kellogg (1969), Ortega–Rheinboldt (1968)]. Let $G(x) = H(x) + V(x)$ with H and V continuous, let $G(\hat{x}) = 0$, let

$$\langle V(x) - V(y), x - y \rangle \geq 0$$

for all x, y in $\mathbb{R}^l$, and for each bounded set B let there exist positive constants L_B and α_B such that

$$\|H(x) - H(y)\| \leq L_B \|x - y\| \quad \text{and}$$
$$\langle H(x) - H(y), x - y \rangle \geq \alpha_B \|x - y\|^2$$

for all x, y in B. Suppose that

$$0 < r \leq s_{n+1} \leq r_n \leq s_n \leq s < \infty$$

for $n \geq 0$ and that

$$\lim_{n \to \infty} r_n = \lim_{n \to \infty} s_n = r$$

Then the sequence $x_{n/2}$ generated by the nonlinear ADI method converges to $\hat{x}$ for every initial x_0.

For the nonlinear Jacobi and SOR methods, one can also compute with monotone iterates.

PROPOSITION 8.3.4 [Ortega–Rheinboldt (1967a, b, 1968)]. Let A be an M-matrix, $\phi: \mathbb{R}^l \longrightarrow \mathbb{R}^l$ be continuous, diagonal, and isotone, and suppose that $y_0 \leq x_0$ satisfies $G(y_0) \leq 0 \leq G(x_0)$ where $G(x) = Ax + \phi(x)$. Then the nonlinear Jacobi and (for $0 < w \leq 1$) nonlinear SOR methods starting from x_0 and y_0 yield sequences satisfying

$$y_n \leq y_{n+1} \leq \hat{x} \leq x_{n+1} \leq x_n \quad \text{for} \quad n \geq 0$$

and $x_n \longrightarrow \hat{x}$, $y_n \longrightarrow \hat{x}$, where $G(\hat{x}) = 0$.

We now look at the Newton–κ methods; the following results are typical.

PROPOSITION 8.3.5 [Greenspan–Parter (1965), Ortega–Rheinboldt (1967a, 1968)]. Let $G: \mathbb{R}^l \longrightarrow \mathbb{R}^l$ be continuously differentiable and convex, let G'_x be an M-matrix for all x, and let $G(\hat{x}) = 0$. Then for any x_0 satisfying $G(x_0) \geq 0$, and for any $m \geq 1$ (actually m may vary) and $0 < w \leq 1$, the iterates generated by the Newton–m-step–SOR method converge to $\hat{x}$ and satisfy $x_n \geq x_{n+1} \geq \hat{x}$ for $n \geq 0$. If $G(x) = Ax + \phi(x)$ where A is an M-matrix and ϕ is continuously differentiable, diagonal, isotone, and convex, then the Newton–1-step–Gauss–Seidel iterates converge to $\hat{x}$ for *any* x_0 and monotonically if $G(x_0) \geq 0$.

PROPOSITION 8.3.6 [Caspar (1968)]. Let $G(x) = H(x) + V(x)$ with H and V continuously differentiable, let $G(x) - G(y) \leq G'_x(x - y)$ if $x \leq y$ or $y \leq x$, let $rI + H'_x$ and $rI + V'_x$ be M-matrices for all $r > 0$ and all x, let $G(\hat{x}) = 0$ and $x_0 \geq \hat{x}$, $G(x_0) \geq 0$. Then, for the Newton–1-step–ADI method, if

$$d(x_n) \leq r_n = s_n \leq s < \infty$$

where $d(x_n)$ is the maximum of the diagonal elements of H'_{x_n} and V'_{x_n}, then the iterates satisfy $x_n \geq x_{n+1} \geq \hat{x}$ and $x_n \longrightarrow \hat{x}$.

For ADI–Newton methods, we have the following result.

PROPOSITION 8.3.7 [Caspar (1968)]. Let H, V, G, H', V', x_0, and x^* satisfy the hypotheses of Proposition 8.3.6. For the ADI–1-step–Newton method with

$$r_n = s_n \leq s < \infty \quad \text{and} \quad r_{n+(1/2)} = s_{n+(1/2)} \leq s < \infty$$

if r_n is greater than or equal to each diagonal element of V'_{x_n} and $r_{n+(1/2)}$ is greater than or equal to each diagonal element of H'_{x_n}, then

$$x_{n/2} \geq x_{(n+1)/2} \geq \hat{x}$$

and $x_{n/2} \longrightarrow \hat{x}$.

To treat SOR–Newton methods, however, it has appeared necessary to use variational methods entirely. Thus one considers the SOR–1-step–Newton method applied at x_n as yielding a direction p_n, and then one must set $x_{n+1} = x_n + t_n p_n$ for suitable t_n just as in the full Newton method analyzed in Theorem 8.2.1. The step t_n can be chosen by essentially any of the usual step-size methods, but we know of no general results allowing $t_n = 1$, even for large n. One of the simplest methods for choosing t_n is one of the techniques of Section 4.6, which in this case is as follows.

PROPOSITION 8.3.8 [Elkin (1968)]. Suppose f''_x is continuous and $0 < aI \leq f''_x$ on $W(x_0)$. Let p_n be the direction generated by the Gauss–Seidel–1-step–Newton method applied at x_n to solve $\nabla f(x) = 0$. Let

$x_{n+1} = x_n + t_n p_n$ where, for $\alpha \in (0, 1)$, t_n is the first of the numbers $t = \alpha^0, \alpha^1, \alpha^2, \ldots$, to satisfy

$$f(x_n) - f(x_n + tp_n) \geq -\delta t_n \langle \nabla f(x_n), p_n \rangle$$

for a fixed $\delta > 0$. Then $x_n \longrightarrow \hat{x}$, minimizing f.

Most other results concerning the choice of t_n restrict t_n to an interval, as in Section 4.5, depending on values of the eigenvalues of $f''_{x'}$; these results as well as that above are valid for block methods also [Elkin (1968), Schechter (1968)].

8.4. LEAST-SQUARES PROBLEMS

One approach for solving the operator equation $J(x) = 0$, and one which in fact is the source of many minimization problems, is that of minimizing $E(x) = \frac{1}{2} ||J(x)||^2$; we shall consider this general question without the assumption that an $\hat{x}$ exists with $J(\hat{x}) = 0$. Thus we shall seek a point $\hat{x}$ where $\nabla E(\hat{x}) = 0$—that is,

$$J'^*_{\hat{x}} J(\hat{x}) = 0$$

where $*$ indicates conjugate transpose. If $J'_{\hat{x}}$ is invertible, this will then imply $J(\hat{x}) = 0$. If we write

$$G(x) = J'^*_x J(x)$$

then we merely have a general nonlinear operator equation $G(x) = 0$ where in fact $G = \nabla E$ and we can use any of the methods mentioned before. For example, we *can* consider for many problems the Newton direction for the equation $J(x) = 0$ and use this as a descent direction for $E(x)$. Since

$$\langle -J'^{-1}_x J(x), \nabla E(x) \rangle = -||J(x)||^2$$

this generates directions driving $||J(x)||$ to zero if $||J'^{-1}_x|| \leq B$ for all x. This approach has been widely used [Fletcher (1968)] with a modification to avoid derivative calculations, so we leave it until the next chapter. We shall now dwell on modifications of Newton's method for this problem. To use Newton's method one needs G'_x; we see here that

$$G'_x = J''_x J(x) + J'^*_x J'_x$$

Thus to use Newton's method one needs second derivatives of J, while J may itself involve first derivatives of some function f. We note, however, that in the case in which we expect $J(\hat{x}) = 0$, the term $J''_x J(x)$ for x near $\hat{x}$

is very small; we might hope, therefore to be able to replace G'_x by $J'^*_x J'_x$. If this latter operator is invertible, we are led to the *Gauss–Newton method* for solving $J'^*_x J(x) = 0$—namely,

$$x_{n+1} = x_n - [J'^*_n J'_n]^{-1} J'^*_n J(x_n)$$

We note that the direction $p_n = -[J'^*_n J'_n]^{-1} J'^*_n J(x_n)$ is a direction of decrease for $E(x)$ since $\langle p_n, \nabla E(x_n)\rangle < 0$ if $J'^*_n J(x_n) \neq 0$. The presentation in Proposition 8.2.1 for Newton-like methods is sufficiently general to provide us with a local-convergence proof for this method; similar general results applied to this case are in Pereyra (1967). For global-convergence properties we can immediately turn to our general results on variational methods if we are willing to modify the method via

$$x_{n+1} = x_n + t_n p_n$$

and allow $t_n \neq 1$. What we really need is for the direction sequence to be admissible, which will be valid if, for example,

$$0 < aI \leq J'^*_x J'_x \leq AI < \infty.$$

EXERCISE. If J'^{-1}_x exists, show that $0 < aI \leq J'^*_x J'_x \leq AI < \infty$ will hold, for example, if there is a $B < \infty$ such that $\|J'_x\| \leq B$, $\|J'^{-1}_x\| \leq B$, for all x.

If $J'^*_x J'_x$ is singular, then the inverse of course makes no sense and the Gauss–Newton iteration is not defined. However, we can consider an extension of the method in which the inverse is replaced by a pseudo-inverse. Denoting the (Penrose) pseudo-inverse of a matrix A by A^+ [Langenhop (1967), Penrose (1955)], we have the iteration

$$x_{n+1} = x_n - [J'^*_n J'_n]^+ J'^*_n J(x_n)$$

This extension of the Gauss–Newton method, along with the Newton method similarly modified, can be analyzed just as the original methods. If, essentially, the pseudo-inverse operation is continuous in some neighborhood of the iterates $\{x_n\}$, then a local-convergence proof can be given for the solution of $[J'^*_x J'_x]^+ J'^*_x J(x) = 0$ [Ben-Israel (1965, 1966)]. Under the same kind of hypotheses, global-convergence results for the usual choices of the step t_n are known.

This type of method appears quite useful for such least-squares problems; we remind the reader, however, that all of the preceding variational techniques are available here also, including the very powerful conjugate-gradient-type methods with the "conjugatizing" performed with respect to $J'^*_x J'_x$—that is, with $H_x \equiv I$ in the general notation for conjugate gradients. One of the drawbacks in practice of the Gauss–Newton approach is that in some cases

$E(x) = \frac{1}{2}\|J(x)\|^2$ cannot be reduced appreciably in the computed direction p_n, often because p_n makes nearly a right angle with $\nabla E(x)$, although such an angle is not necessarily bad. In such a case it is sometimes wise to use the gradient direction instead, or, perhaps better yet, some compromise between the pure gradient and pure Gauss–Newton directions. We consider this approach next.

One way [Kowalik–Osborne (1969), Levenberg (1944), Marquardt (1963), Morrison (1960)] of effecting such a compromise between the directions is to use the direction $p_n(\alpha_n)$ defined by

$$[J_n'^*J_n' + \alpha_n I]p_n = -J_n'^*J(x_n), \ \alpha_n \geq 0$$

For $\alpha_n = 0$, this yields the Gauss–Newton direction, while for α_n large, the direction is

$$p_n = \frac{1}{\alpha_n}\nabla E(x_n) + O\left(\frac{1}{\alpha_n^2}\right)$$

which is asymptotically the gradient direction.

EXERCISE. Show that, as α_n increases, $\|p_n(\alpha_n)\|$ and the angle between $p_n(\alpha_n)$ and $-\nabla E(x_n)$ decrease monotonically to zero.

Since, moreover, $J_n'^*J_n'$ is positive-semidefinite, we can always solve for p_n if $\alpha_n > 0$ and, if α_n is bounded away from zero, $[J_n'^*J_n' + \alpha_n I]$ will be uniformly positive-definite; therefore, if $\|J_n'\|$ is bounded and $0 < a \leq \alpha_n \leq A < \infty$, then the direction sequence is admissible and the various methods for choosing t_n may be used. In practice, of course, one of the difficulties is in choosing α_n so as to get a significant decrease in $E(x)$ without moving too far from the Gauss–Newton direction, which should, if $J(x^*) = 0$, yield ultimately superlinear convergence. Since computing $p_n(\alpha_n)$ involves solution of a linear system for each value of α_n, one apparently would not want to experiment with different values of α_n for fixed n if this can be avoided; in fact this can be done easily using matrix decompositions [Golub-Saunders (1969)].

A more direct way of effecting a compromise between two choices of direction has recently been proposed [Powell (1968, 1969)]. The method there is merely to set

$$-p_n(\alpha_n) = \alpha_n J_n'^{-1}J(x_n) + (1 - \alpha_n)J_n'^*J(x_n), \qquad \alpha_n \in [0, 1]$$

the compromise here being between $\nabla E(x_n)$ and the *pure Newton direction* for J, although the Gauss–Newton direction could be used as well. Because each of the directions of which $p_n(\alpha_n)$ is a convex combination is itself a direction of decrease for $E(x)$, so also is $p_n(\alpha_n)$. In fact, if α_n is bounded away from 1, even if the inverse is replaced by the pseudo-inverse the directions

$p_n(\alpha_n)$ will be admissible and our various step-size algorithms are applicable. Again, the problem of choosing α_n at each step is not well understood; however, trying various choices is not too costly, since a set of linear equations is only solved once. A heuristic approach in Powell (1968) is used to select α_n, with excellent results; since the method there is actually implemented with differences—as opposed to derivatives—and with other computationally convenient modifications, we postpone further discussion to the next chapter.

9 AVOIDING CALCULATION OF DERIVATIVES

9.1. INTRODUCTION

In the preceding chapters a variety of methods has been presented for minimizing a functional f; these methods require a variety of kinds of data about f, such as values of f, ∇f, and f''. Although for certain important problems the derivatives needed in ∇f and f'' are easy to compute, generally speaking we should like to avoid using any more derivatives than absolutely necessary. Thus recently, methods have been created to reduce dependency on knowledge of derivatives. In some cases these methods take the simple expedient of approximating the derivatives, usually by differences, but there have also been some techniques developed which entirely avoid all derivatives and approximation thereto. In this chapter we shall examine some of these methods and see how they can be analyzed.

9.2. MODIFYING DAVIDON'S FIRST METHOD

Recall that the first method of Davidon is a variable-metric algorithm which proceeds as follows, given x_0 and H_0:

Given x_n and H_n, set

$$p_n = -H_n \nabla f(x_n), \qquad x_{n+1} = x_n + t_n p_n$$

where t_n minimizes $f(x_n + tp_n)$; let $\delta_n = \nabla f(x_{n+1}) - \nabla f(x_n)$ and let

$$H_{n+1} = H_n + t_n \frac{p_n p_n^*}{\langle p_n, \delta_n \rangle} - \frac{H_n \delta_n \delta_n^* H_n}{\langle \delta_n, H_n \delta_n \rangle}$$

where * indicates conjugate transpose. This method requires use of the derivatives $\nabla f(x_n)$; these can be approximated very simply by the vector γ of forward differences—that is,

$$\gamma_i = \frac{f(x_n + d_i e_i) - f(x_n)}{d_i}, \qquad d_i > 0$$

where e_i is the vector with $e_{ij} = \delta_{ij}$, the Kronecker delta. It is clear that if the d_i for $i = 1, 2, \ldots, l$ are small enough, this method will behave essentially as well as the method with derivatives; the problem here, as with all the other methods wherein derivatives are replaced by differences, lies in how to choose the d_i. An excellent analysis of this problem has been given for Davidon's method [Stewart (1967)]; since the viewpoint is of interest for use on any method, we present the ideas here once and for all.

The whole basis of most gradient methods is to treat $f(x)$ as a quadratic, locally; thus we shall consider the problem of approximating the derivative $\gamma = f'(0)$ of the quadratic

$$f(t) = f(0) + \gamma t + \tfrac{1}{2}\alpha t^2$$

by the difference

$$\frac{f(d) - f(0)}{d} \equiv \gamma_d$$

The two sources of error in approximating the scalar γ by the scalar γ_d are the truncation error in the divided difference and the cancellation produced in computing $f(d) - f(0)$ for small d; clearly we should balance these errors. We can estimate the relative truncation error by

$$\left|\frac{\gamma_d - \gamma}{\gamma}\right| \doteq \frac{1}{2}\left|\frac{\alpha d}{\gamma}\right|$$

If we assume that, in computing $f(t)$, we *actually* compute

$$f^*(t) = f(t)(1 + \epsilon), \qquad |\epsilon| \leq \eta \quad \text{known}$$

then the relative cancellation error can be estimated as

$$2\left|\frac{f(0)}{f(d) - f(0)}\right|\eta$$

If we equate the two estimates and solve for the number d, we find that $|d|$ should be the positive root of

$$\tfrac{1}{2}\alpha^2 z^3 + |\alpha||\gamma| z^2 - 4|f(0)||\gamma|\eta = 0$$

and sign (d) = sign $(\alpha\gamma)$.

EXERCISE. Show that the above choice of d is correct.

To avoid solving the above cubic, experimentally it has been satisfactory to ignore the cubic or quadratic term, depending on which gives a smaller root, to solve the resulting simpler equation, and to refine the result by one Newton step applied to the original cubic. This gives the following:

$$\tau = 2\left\{\left|\frac{f(0)}{\alpha}\right|\right\}^{1/2}, \quad |d| = \tau\left[1 - \frac{|\alpha|\tau}{3|\alpha|\tau + 4|\gamma|}\right] \quad \text{if} \quad \gamma^2 \geq |\alpha f(0)|\eta$$

$$\tau = 2\left\{\left|\frac{f(0)}{\alpha}\right|\right\}^{1/3}, \quad |d| = \tau\left[1 - \frac{2|\gamma|}{3|\alpha|\tau + 4|\gamma|}\right] \quad \text{if} \quad \gamma^2 < |\alpha f(0)|\eta$$

This computation requires a crude estimate of γ, which can easily be obtained, and also an estimate α of $f''(0)$. For the Davidon method we are considering,

$$f(x_n + te_i) = f(x_n) + \langle \nabla f(x_n), e_i \rangle t + \tfrac{1}{2}\alpha_{ii}t^2 + o(t^2)$$

where α_{ii} is the ith diagonal element of f''_n. Recall that $H_n \sim f''^{-1}_n$; thus we seek the diagonal elements of H_n^{-1}. These can in fact be computed recursively (see *Proof* of Theorem 7.4.4) without knowledge of the off-diagonal elements H_n^{-1} from the equations [Stewart (1967)]

$$H_{i+1}^{-1} = H_i^{-1} + \left(\frac{1}{\beta_i t_i} - \frac{\rho_i}{\beta_i^2}\right)\delta_i\delta_i^* + \frac{1}{\beta_i}[\nabla f(x_i)\delta_i^* + \delta_i \nabla f(x_i)^*]$$

where

$$\rho_i = \langle \nabla f(x_i), p_i \rangle, \qquad \beta_i = \langle \delta_i, p_i \rangle$$

and we replace $\nabla f(x_i)$ by its approximation.

EXERCISE. Show that the above recursion for H_{i+1}^{-1} is correct.

Thus we have a rule for determining the size of the numbers d_i in the difference approximation to $\nabla f(x_n)$. Certainly we have ignored many problems, implying that our analysis is far from rigorous, but the ideas in practice appear to lead to good results. For further computational details and examples the reader is referred to Stewart (1967), where the method is shown to be quite powerful in practice. A similar approach has been applied for a Davidon-like method applied to minimize $\|J(x)\|^2$ for a nonlinear operator J [Fletcher (1968)]; here $J_x'^{-1}$ usually does not exist, so one is led to finding a Davidon-type approximation to a pseudo-inverse $J_x'^{+}$ using differences. The method as proposed in Fletcher (1968) is exact for quadratics.

EXAMPLE [Stewart (1967)]. The Davidon modification without derivatives was used to minimize

$$f(x, y) = 100(y - x^2)^2 + (1 - x)^2$$

starting with $x_0 = -1.2$, $y_0 = 1.0$. After 163 function evaluations, f was reduced to 9×10^{-12} with

$$(x, y) = (1.000002, 1.000003)$$

9.3. MODIFYING NEWTON'S METHOD

Computationally, at least two problems are involved in Newton's method: the need for solving a linear system at each step, and the need for evaluating roughly l^2 derivatives at each step. One can eliminate most of the derivative evaluation by using the derivatives at one fixed point throughout, but this eliminates the powerful feature of quadratic convergence. An alternative, of course, is to use differences to evaluate the derivatives. If the step size used for the differences is small enough, one should maintain the rapid convergence, it would seem. The local-convergence properties can be analyzed by means of Proposition 8.2.1; if we replace the derivative of $J(x)$ by $\Delta J(x, \epsilon) \equiv \gamma$, where the components

$$\gamma_i = \frac{1}{\epsilon}[J(x_n + \epsilon e_i) - J(x_n)]$$

as in Section 9.2, we have the following local result [Dennis (1969)].

PROPOSITION 9.3.1. Let

$$A_0^{-1} \equiv [\Delta J(x_0, \epsilon_0)]^{-1}$$

exist with

$$\|A_0^{-1}J'_x - A_0^{-1}J'_y\| \leq K\|x - y\|$$

for $x, y \in S(x_0, r)$. Let $\epsilon > 0$ be such that

$$K\left(\epsilon + \frac{1}{2}\epsilon_0\right) \leq 1 \quad \text{and} \quad \frac{1}{2} > h = \frac{K\|A_0^{-1}J(x_0)\|}{(1 - K\epsilon - \frac{1}{2}K\epsilon_0)^2}$$

Suppose that

$$r \geq r_0 \equiv \frac{1 - \sqrt{1 - 2h}}{h} \times \frac{\|A_0^{-1}J(x_0)\|}{1 - K\epsilon - \frac{1}{2}K\epsilon_0}$$

and that ϵ_n is a sequence of numbers such that $|\epsilon_n| \leq \epsilon$ and $x_n + \epsilon_n e_i \in S(x_0, r)$ for $i = 1, \ldots, l$. Then the sequence

$$x_{n+1} = x_n - [\Delta J(x_n, \epsilon_n)]^{-1}J(x_n)$$

is well defined and converges to $\hat{x}$, solving $J(x) = 0$. If, for a constant C, we have $|\epsilon_n| \leq C||J(x_n)||$, then the convergence is quadratic.

For the global properties the situation is somewhat less simple; we have to be sure that the directions generated are descent directions. Thus let us suppose that $J(x) = \nabla f(x)$ and that $\Delta J(x, \epsilon)$ is as defined before; let us consider the follwoing direction-generating algorithm [Goldstein–Price (1967)]:

If $\Delta J(x_n, \epsilon_n)$ is singular, or if

$$\langle \nabla f(x_n), [\Delta J(x_n, \epsilon_n)]^{-1} \nabla f(x_n) \rangle \leq 0$$

then set $p_n \equiv p_n(x_n) = -\nabla f(x_n)$; otherwise set (9.3.1)

$$p_n = -[\Delta J(x_n, \epsilon_n)]^{-1} \nabla f(x_n)$$

Suppose that $0 < aI \leq J'_x \leq AI$; clearly this "uniform nonsingularity" of J'_x implies, if J'_x is continuous in x, that for bounded sequences x_n with $\epsilon_n \longrightarrow 0$, we have that $||\Delta J(x_n, \epsilon_n) - J'_n|| \longrightarrow 0$, that $\Delta J(x_n, \epsilon_n)$ is nonsingular for large n and is in fact positive-definite, and that

$$||[\Delta J(x_n, \epsilon_n)]^{-1} - J'^{-1}_n|| \longrightarrow 0$$

Thus we may use, as usual, any of the step-determining methods in conjunction with the admissible direction sequence this generates. In particular, if we can take $t_n = 1$ for large n and $|\epsilon_n| \leq C||\nabla f(x_n)||$, we would expect quadratic convergence near the solution $\hat{x}$ because of Proposition 9.3.1 if J'_x is Lipschitz-continuous in x near $\hat{x}$. The following theorem describes how this can be done by means of the method of Section 4.6.

THEOREM 9.3.1. Let $0 < aI \leq J'_x \leq AI$ for all x, with $J(x) = \nabla f(x)$ and J'_x continuous in x. Let the directions p_n for $n \geq 1$ be generated by Equation 9.3.1 with $|\epsilon_n| \leq C||t_{n-1}p_{n-1}||$ and let $p_0 = -\nabla f(x_0)$. Let $q \in (0, \frac{1}{2})$ and $\alpha \in (0, 1)$ be given and choose t_n as the first of the numbers $t = \alpha^0, \alpha^1, \alpha^2, \ldots,$ for which

$$f(x_n) - f(x_n + tp_n) \geq qt\langle -\nabla f(x_n), p_n \rangle$$

and let $x_{n+1} = x_n + t_n p_n$. Then for any x_0, the sequence $\{x_n\}$ converges to the unique $\hat{x}$ minimizing f over $\mathbb{R}^l$. There exists an N such that for $n \geq N$ we have $t_n = 1$; if J'_x is Lipschitz-continuous in some neighborhood of $\hat{x}$, then $x_n \longrightarrow \hat{x}$ quadratically.

Proof: The iteration is certainly well defined, since $\langle -\nabla f(x_n), p_n \rangle > 0$ unless $x_n = \hat{x}$. Therefore, we have

$$qt_n\langle -\nabla f(x_n), p_n\rangle \leq f(x_n) - f(x_n + t_n p_n)$$
$$= -t_n\langle -\nabla f(x_n), p_n\rangle - \frac{t_n^2}{2}\langle J'_{x_n+\lambda_n t_n p_n} p_n, p_n\rangle$$
$$\text{for some } \lambda_n \in (0, 1)$$

and hence

$$(1-q)\langle -\nabla f(x_n), t_n p_n\rangle \geq \frac{a}{2}\, \| t_n p_n \|^2$$

Since f is bounded below and

$$f(x_n) - f(x_{n+1}) \geq q\langle -\nabla f(x_n), t_n p_n\rangle$$

it follows that $\langle -\nabla f(x_n), t_n p_n\rangle$—and hence $\| t_n p_n \|^2$—tends to zero. Therefore, $\epsilon_n \longrightarrow 0$ and, since $f(x_n) \leq f(x_0)$, the sequence $\{x_n\}$ is bounded; hence

$$\| \Delta J(x_n, \epsilon_n) - J'_n \| \longrightarrow 0, \qquad \| [\Delta J(x_n, \epsilon_n)]^{-1} - J_n'^{-1} \| \longrightarrow 0$$

Hence for large n we have

$$\frac{a}{2} I \leq \Delta J(x_n, \epsilon_n) \leq 2AI$$

We may now apply Corollary 4.6.2 with

$$d(t) = qt, \qquad d_1(t) = t \min\left(\frac{1}{2A}, 1\right)$$

and conclude that $\nabla f(x_n) \longrightarrow 0$. Thus, by Theorem 4.2.1 and either Proposition 1.6.1 or Theorem 1.6.1, we deduce that $x_n \longrightarrow \hat{x}$. Arguing just as in the *Proof* of Theorem 8.2.1 and using the fact that

$$J_n'^{-1} = [\Delta J(x_n, \epsilon_n])^{-1} + o(1) \quad \text{as} \quad n \longrightarrow \infty$$

we conclude that we may take $t_n = 1$ for large n. If J'_x is Lipschitz-continuous in x, we conclude from Proposition 9.3.1 that the convergence is quadratic. Q.E.D.

In practice this method appears [Goldstein–Price (1967)] to exhibit rapid convergence without the cost in time caused by derivative evaluation.

EXAMPLE [Goldstein–Price (1967)]. The method was applied to minimize $f(x, y) = 100(y - x^2)^2 + (1 - x)^2$ and generally required about one-third as many function and gradient evaluations as the original Davidon method to reduce the function to a given value. In particular, about 70 and

205 evaluations were required by the respective methods to reduce f from an initial value of about 24 to a value of about 10^{-14}.

A similar modification of the Gauss–Newton method that has been proposed also attempts to reduce computing time by eliminating some of the difficulties caused by having to solve a linear system at each step of the iteration; the technique is certainly applicable to the general Newton method, but we shall discuss it for the Gauss–Newton method, to which it was first applied.

9.4. MODIFYING THE GAUSS–NEWTON METHOD

Recall that the Gauss–Newton method is applied to minimize $E(x) \equiv \frac{1}{2}||J(x)||^2$ for a nonlinear operator J mapping $\mathbb{R}^l$ into $\mathbb{R}^m$; the iteration proceeds by setting $x_{n+1} = x_n + t_n p_n$, where t_n minimizes $E(x_n + tp_n)$ and $p_n = -(J_n'^* J_n')^{-1} J_n'^* J(x_n)$ where * indicates conjugate transpose. We wish to eliminate the derivative calculations and reduce the difficulty inherent in computing p_n as the solution of a linear system at each iteration. The theory behind approximating the derivatives causes no problem, as we saw in the last section; we consider the following approach for treating the other difficulty, essentially as in Powell (1964a).

Suppose that at the iterate x_n we have l (in $\mathbb{R}^l$) independent directions $d_{n,1}, d_{n,2}, \ldots, d_{n,l}$ and estimates $\gamma_{n,i}$ of the derivative $J_n'^* J_n' d_{n,i}$ of E at x_n in the direction $d_{n,i}$ for $i = 1, 2, \ldots, l$. If we write $p_n = \sum_{i=1}^{l} q_{n,i} d_{n,i}$ for unknown scalars $q_{n,i}$, then we need to solve

$$J_n'^* J_n' p_n = -J_n'^* J(x_n),$$

that is,

$$J_n'^* J_n' \left[\sum_{i=1}^{l} q_{n,i} d_{n,i} \right] \equiv \sum_{i=1}^{l} q_{n,i} \gamma_{n,i} = -J_n'^* J(x_n)$$

Assuming, of course, that J_n' is itself approximated by some matrix P_n, probably by differences, we must solve the linear system

$$\Gamma_n q_n = -P_n^* J(x_n)$$

where

$$\Gamma_n = (\gamma_{n,1}, \gamma_{n,2}, \ldots, \gamma_{n,l}), \qquad q_n = (q_{n,1}, \ldots, q_{n,l})$$

During the minimization of $E(x_n + tp_n)$ it is quite simple to compute estimates of the derivative $J_n'^* J_n' p_n$ of E along p_n; therefore, it is natural to let p_n replace one of the directions $d_{n,i}$. With some extra effort one can choose the d_{n,i_0}

to be replaced in such a way that the new direction set is as "linearly independent" as possible [Powell (1964a)]; that direction and its derivative estimate are then replaced, while the other directions *and derivative estimates* are kept fixed. This is based on the assumption that $J_{n+1}'^{*}J_{n+1}'d_{n,i}$ is not significantly different from $J_n'^{*}J_n'd_{n,i}$, or, if it is different, that the direction $d_{n,i}$ and its derivative estimate will soon be replaced automatically. Thus, from step n to step $(n+1)$, only one vector $\gamma_{n,i}$ is changed—that is, only one column of Γ_{n+1} is different from that of Γ_n; in this situation, however, Γ_{n+1}^{-1} can be computed by a simple updating of Γ_n^{-1}. Therefore, if l is small enough that the $l \times l$ matrices Γ_n^{-1} can be stored, the iteration can proceed rapidly without constantly resolving a linear system. Clearly a local-convergence proof for this method can be given under suitable hypotheses; we are not aware of any global-convergence results, however. At any rate, this heuristic approach and other modifications of it appear quite useful.

EXAMPLE [Powell (1964a)]. The method of Powell (1964a) was applied to minimize

$$f(x, y) = [10(y - x^2)]^2 + [(1 - x)]^2$$

starting at $x_0 = -1.2$, $y_0 = 1.0$. About 65 function evaluations yielded $(x, y) = (0.9919, 0.9827)$ with $f = 0.0002$, while about 70 yielded $(x, y) = (0.9986, 0.9973)$ with $f = 0.000003$.

9.5. MODIFYING THE GAUSS-NEWTON-GRADIENT COMPROMISE

In Section 8.4 we considered some "compromise-direction" methods, one given by

$$(J_n'^{*}J_n' + \alpha_n I)p_n = -J_n'^{*}J(x_n)$$

where * indicates conjugate transpose, and another given by

$$p_n = \alpha_n N_n + (1 - \alpha_n)G_n, \qquad \alpha_n \in [0, 1]$$

where N_n and G_n were two descent directions; it is proposed in Powell (1968) that

$$G_n = -\nabla E(x_n) = -J_n'^{*}J(x_n) \quad \text{and} \quad N_n = -J_n'^{-1}J(x_n)$$

although $N_n = -[J_n'^{*}J_n']^{-1}J_n'^{*}J(x_n)$ might also be used, especially if $J\colon \mathbb{R}^l \longrightarrow \mathbb{R}^m$ and J_x' is not invertible. Computationally, the more direct method of the two, setting

$$p_n = \alpha_n N_n + (1 - \alpha_n)G_n$$

appears more reasonable and we therefore discuss the implementation of this approach. The algorithm as presented in Powell (1968) gives a full description of an operational method of apparently great power, partially because of the way in which it attempts to simplify computations by using some of the various techniques described in other sections and chapters of this text; we sketch the approach to show the way in which many concepts can be combined.

First of all, the computation of N_n as first described requires solution of a linear system. This can be handled instead by a variable-metric approach with $N_n = -H_nJ(x_n)$ [or $-H_nJ_n'^*J(x_n)$] where H_n is computed recursively as an approximation to the appropriate matrix inverse; we saw in Sections 7.2 and 7.3 how simply H_n could be generated to satisfy this condition, and in Section 9.2 how all derivatives could be replaced by differences. Once N_n and G_n are available, we try to generate p_n. In order to avoid searching for the minimum of $E(x)$ along $x_n + tp_n$ each time, the computer program in Powell (1968) carries along a varying estimate s_n of the step $||t_np_n||$, computed by using essentially the previous step $||t_{n-1}p_{n-1}||$. Therefore, if $||N_n|| \leq s_n$, the direction $p_n = N_n$ is used; if $||N_n|| > s_n$, some component of G_n is introduced. Now the minimum of $E(x)$ along $x_n + tG_n$ can be estimated to be at

$$x_n + \mu G_n, \qquad \mu \equiv \frac{||G_n||^2}{||J_n'G_n||^2}$$

If $s_n \leq \mu ||G_n||$, the direction vector p_n is chosen as

$$p_n = s_n \frac{G_n}{||G_n||}$$

Otherwise we set

$$p_n = \alpha_n N_n + (1 - \alpha_n)G_n\mu$$

with α_n chosen so that $||p_n|| = s_n$. If $E(x_n + p_n)$ is "sufficiently" smaller than $E(x_n)$, then x_{n+1} is set equal to $x_n + p_n$, while otherwise s_n is "reduced" and the direction-generating process recommences; the terms "sufficiently" and "reduced" are defined in Powell (1968) heuristically in an attempt to guarantee convergence without having an excessively small step size $||x_{n+1} - x_n||$ very often.

EXAMPLE [Powell (1968)]. The implementation of the above method in Powell (1968) was applied to minimize

$$f(x, y) = [10(y - x^2)]^2 + [(1 - x)]^2$$

starting with $x_0 = -1.2$, $y_0 = 1.0$. After 26 function evaluations, the point $(x, y) = (1.0000, 1.0014)$ had been reached with $f = 0.0002$.

The rough outline of the above computational algorithm was given only to show how various computational methods could be blended together to avoid such problems as derivative computation, matrix inversion, and so forth; it is certainly not possible for us to list all possible techniques, but the reader should now have a basic grasp of the *kinds* of techniques available for simplifying computations and of the variety of ways in which they can be combined. Evidence seems to indicate that, for efficiency without loss of accuracy, those methods which avoid derivative calculations by using differences and avoid the matrix inversion of Newton or Gauss–Newton methods by using variationally oriented variable-metric Davidon-type methods are the best available at present.

9.6. METHODS IGNORING DERIVATIVES

The methods mentioned so far in this chapter have been simply *modifications* of other methods in order to avoid the calculation of derivatives required by those methods. We turn now to methods, designed essentially for non-smooth or noisy functions, which do not, even by approximation, attempt to make any use of derivatives. Perhaps the earliest such method is the *basic relaxation method* [Southwell (1946)], in which the functional $f(x)$, $x \in \mathbb{R}^l$, is minimized as a function of each coordinate direction e_i successively—that is, $x_{n+1} = x_n + t_n p_n$, where t_n minimizes $f(x_n + tp_n)$, and $p_n = e_i$, where $i = n(\text{modulo } l) + 1$; sometimes also t_n was chosen merely to decrease $f(x_n)$ sufficiently.

Since the above method does not allow for diagonal movement, the following method was developed [Hooke–Jeeves (1961)]. Staring at any x, we define the *exploratory-move operator* $\text{EM}(x)$ as follows. Given some step sizes $\delta_i = \delta_i(x) > 0$, setting $x' = x$, starting with $i = 1$ up to $i = l$, if $f(x' + \delta_i e_i) < f(x')$, then x' is replaced by $x' + \delta_i e_i$ and i by $i + 1$; if $f(x' + \delta_i e_i) \geq f(x')$, but $f(x' - 2\delta_i e_i) < f(x')$, then x' is replaced by $x' - 2\delta_i e_i$ and i by $i + 1$; otherwise x' is not changed. Finally, when i reaches $l + 1$, we set $\text{EM}(x) \equiv x'$. The entire algorithm proceeds from x_n to x_{n+1} as follows, starting with some initial x_0 and $x_1 \equiv \text{EM}(x_0) \neq x_0$. We compute $x'_n = \text{EM}(x_n)$; if $x'_n = x_n$, then $\delta_i(x_n)$ is cut in half for $i = 1, 2, \ldots, l$ and the iteration restarts at x_n. Otherwise, if $f[\text{EM}(2x_n - x_{n-1})] < f(x_n)$, we set $x_{n+1} = \text{EM}(2x_n - x_{n-1})$; if the latter inequality is invalid, we set $x_{n+1} = x'_n = \text{EM}(x_n)$. If f is strictly convex with ∇f continuous and $\lim_{\|x\| \to \infty} f(x) = +\infty$, then it can be shown [Céa (1969)] that $\|x_n - \hat{x}\| \to 0$, where $\hat{x}$ minimizes f over $\mathbb{R}^l$. A computer implementing this method can be found in Kaupe (1963).

Since the coordinate directions, which are used in the above algorithm, need not be the best ones, the process has been modified as follows [Rosenbrock (1960)]. Given a vector x and l orthonormal directions $d_i(x)$, l step sizes $\delta_i(x)$, and two parameters $\alpha > 1$, $\beta \in (0, 1)$, the *exploratory-move operator* EM(x) is defined as follows. For i cycling through the values $1, 2, \ldots, l$, setting $x' = x$, if $f(x' + \delta_i d_i) < f(x')$, then we replace x' by $x' + \delta_i d_i$, i by $i + 1$ (or l by 1), δ_i by $\alpha\delta_i$, and record a *success;* otherwise δ_i is replaced by $-\beta\delta_i$ and a *failure* is recorded. After one success and one failure have been recorded for each value of i, we set EM(x) $= x'$. The iteration, starting with x_0 and directions $d_1(x_0), \ldots, d_l(x_0)$, now proceeds from x_n to x_{n+1} and from $d_i(x_n)$ to $d_i(x_{n+1})$ as follows. Set $x_{n+1} = \text{EM}(x_n)$. Let λ_i be the sum of the steps in the direction $d_i(x_n)$ and define vectors $q_i(x_n) \equiv \sum_{j=1}^{l} \lambda_j d_j(x_n)$. New directions $d_i(x_{n+1})$ are now obtained by orthonormalizing the vectors q_i; this completes the description of the method. Roughly speaking, we can say that $d_1(x_{n+1})$ is the most successful motion found so far, $d_2(x_{n+1})$ is the most successful direction orthogonal to $d_1(x_{n+1})$, and so on. A further modification of this method [Swann (1964)] is, for each direction $d_i(x_n)$, to move to the point minimizing f in that direction, and then to compute new directions as before. We do not consider these methods further since we believe the methods to be considered next to be of greater importance and usefulness.

We have seen several times in earlier chapters that there is great advantage to using conjugate directions of some type; the above methods, however, all deal with orthogonal directions—that is, I-conjugate directions. It is possible, however, to generate directions that are conjugate (at least for quadratics) without dealing with derivatives. These methods, which appear to be the best of those that ignore derivatives, are based on the fact that if one minimizes a quadratic

$$f(x) = \langle h - x, M(h - x)\rangle$$

in the direction p from two points x_1 and x_2, arriving at the points x_1' and x_2', then $x_1' - x_2'$ is M-conjugate to p, since

$$\langle p, M(h - x_1')\rangle = \langle p, M(h - x_2')\rangle = 0$$

implies $\langle p, M(x_1' - x_2')\rangle = 0$.

The first good method to make use of this was given by Smith (1962); the method was considerably simplified without losing the conjugate-gradient property in Powell (1964b), but unfortunately the latter's algorithm could break down even for quadratics [Zangwill (1967)]. A correct version of this idea was developed in Zangwill (1967), and it is this version we shall present. The method is as follows.

Let the initial directions $d_{0,i}$, $i = 1, \ldots, l$ be the coordinate directions, $d_{0,i} \equiv e_i$, let $e \in (0, 1]$, set $\delta_1 = 1$, and let $x_{1,0}$ be the initial point. Iteratively apply the *basic k-iteration* starting with $k = 1$; the basic k-iteration is as follows: (1) for $r = 1, 2, \ldots, l$, compute $t_{k,r}$ to minimize $f(x_{k,r-1} + td_{k-1,r})$ and let

$$x_{k,r} = x_{k,r-1} + t_{k,r}d_{k-1,r}$$

(2) Let

$$\alpha_k = \|x_{k,l} - x_{k,0}\| \quad \text{and} \quad d_{k,l+1} = \frac{x_{k,l} - x_{k,0}}{\alpha_k}$$

Compute $t_{k,l+1}$ to minimize

$$f(x_{k,l} + td_{k,l+1})$$

and set

$$x_{k+1,0} = x_{k,l+1} = x_{k,l} + t_{k,l+1}d_{k,l+1}$$

(3) Let $t_{k,s} = \max\{t_{k,r}; r = 1, 2, \ldots, l\}$; if

$$\frac{t_{k,s}\delta_k}{\alpha_k} \geq \epsilon$$

let

$$d_{k+1,r} = d_{k,r} \quad \text{for} \quad r \neq s, \qquad d_{k+1,s} = d_{k,l+1}$$

and let

$$\delta_{k+1} = \frac{t_{k,s}\delta_k}{\alpha_k}$$

If, however,

$$\frac{t_{k,s}\delta_k}{\alpha_k} < \epsilon$$

let $d_{k+1,r} = d_{k,r}$ for $r = 1, 2, \ldots, l$ and set $\delta_{k+1} = \delta_k$.

Consider the method applied to minimize the quadratic

$$f(x) = \langle h - x, M(h - x)\rangle$$

Suppose each time that the last direction $d_{k,l}$ is replaced by $d_{k,l+1}$. Then the last step of the k-iteration and that of the $(k+1)$-iteration are in the same direction and, therefore, because of our earlier remark, we shall next introduce a conjugate direction. After $l + 1$ steps we would have l conjugate directions and, *if* they are linearly independent, we shall therefore get the correct solution on the next iteration. The method of choosing the direction to be eliminated is the technique that keeps the directions $d_{k,1}, \ldots, d_{k,l}$ independent, as we shall see; it also determines which direction, if any, is eliminated at each step and thereby invalidates the above argument. Thus it does not appear possible to prove that this method is *exact* for quadratics, although we can prove convergence. First we show that the directions $d_{k,i}$ (for arbitrary functionals) are linearly independent.

THEOREM 9.6.1 [Zangwill (1967)]. The directions $d_{k,1}, \ldots, d_{k,l}$ are linearly independent. In fact, their determinant satisfies $\det[(d_{k,1}, \ldots, d_{k,l})] = \delta_k \geq \epsilon$.

Proof: The result is true for $k = 1$; assume it for k. Then, since $x_{k,l} - x_{k,0} = \alpha_k d_{k,l+1}$, we have

$$\det[(d_{k,1}, \ldots, d_{k,s-1}, d_{k,l+1}, d_{k,s+1}, \ldots, d_{k,l})]$$
$$= \frac{t_{k,s}}{\alpha_k} \det[(d_{k,1}, \ldots, d_{k,l})] = \frac{t_{k,s}\delta_k}{\alpha_k} \quad \text{for all } s$$

The choice of s—that is, the direction to try to replace—gives us the greatest chance of replacement, while the criterion for replacing or not yields $\delta_{k+1} \geq \epsilon$. Q.E.D.

Having the above fact, we can prove convergence.

THEOREM 9.6.2. Let f be a continuous and strongly quasi-convex functional on $\mathbb{R}^l$, and let the above method be applied starting with an arbitrary $x_{1,0}$; suppose the sequence

$$\{x_{k,r}\}, \qquad r = 0, 1, \ldots, l, \qquad k = 1, 2, \ldots,$$

is bounded. Then any limit point x' of x_{k,r_0} as $k \longrightarrow \infty$ for any $r_0 = 0, 1, \ldots, l$ is also a limit point of $x_{k,r}$, $r \neq r_0$, as $k \longrightarrow \infty$, and for each such limit point there exist l linearly independent directions $d_1, \ldots, d_l$ such that $f(x') \leq f(x' + td_r)$ for all t and $r = 1, 2, \ldots, l$.

Proof: Since $||d_{k,r}|| = 1$ also, given any subsequence K of integers k, we can find a further subsequence K_1 such that $d_{k,r} \longrightarrow d_r$ for $r = 1, 2, \ldots, l$ as $k \longrightarrow \infty$ with $k \in K_1$, and $x_{k,r} \longrightarrow x_r$ for $r = 0, 1, \ldots, l$ as $k \longrightarrow \infty$ with $k \in K_1$; we show next that $x_{r+1} = x_r$ for $r = 0, 1, \ldots, l-1$. Recalling that $f(x_{k,r+1}) \leq f(x_k)$ for all k and $r = 0, 1, \ldots, l-1$ and that $f(x_{k+1,0}) \leq f(x_{k,l})$, we have

$$f(x_{r+1}) = \lim_{k \in K_1} f(x_{k,r+1}) \leq \lim_{k \in K_1} f(x_{k,r}) = f(x_r)$$

In fact, for all t,

$$f(x_{k,r+1}) \leq f(x_{k,r} + td_{k-1,r+1})$$

which then yields, just as above,

$$f(x_{r+1}) = f(x_r) \leq f(x_r + td_{r+1}) \quad \text{for all } t$$

But clearly

$$x_{r+1} - x_r = t_{r+1} d_{r+1}$$

where in fact $t_{r+1} = \lim_{k \in K_1} t_{k,r+1}$, which implies that f is minimized along the line $x_r + td_{r+1}$ at two points x_r and x_{r+1}; because of the convexity assumption on f, we must have $x_r = x_{r+1}$. Denote the common limit by x'—that is,

$$x' = x_0 = x_1 = \cdots = x_l$$

We now have $f(x') \leq f(x' + td_r)$ for all t and $r = 1, \ldots, l$, since this was true for x_{r-1}, as we saw above. However, since

$$\det [(d_{k,1}, d_{k,2}, \ldots, d_{k,l})] \geq \epsilon$$

and $\| d_{k,r} \| = 1$, we must have

$$\det [(d_1, d_2, \ldots, d_l)] \geq \epsilon$$

which implies that the d_i are linearly independent. Since for *any* subsequence K, we could find a further subsequence K_1 as above, the theorem is proved. Q.E.D.

COROLLARY 9.6.1. If, in addition to the hypotheses of Theorem 9.6.2, f is continuously differentiable, then $\nabla f(x') = 0$ for any limit point x' of $x_{k,r}$.

Proof: As we saw above, we have $f(x') \leq f(x' + td_r)$ for all t and $r = 1, \ldots, l$; thus $\langle \nabla f(x'), d_r \rangle = 0$. Since the d_r are linearly independent, we have $\nabla f(x') = 0$. Q.E.D.

COROLLARY 9.6.2. If, in addition to the hypotheses of Corollary 9.6.1, we know that $\{x; \nabla f(x) = 0\}$ contains no continuum, then the sequence $x_{k,r}$ converges.

Proof: We have shown that the difference of successive elements in the sequence $x_{k,0}, \ldots, x_{k,l}$ tends to zero; the same argument shows that

$$\| x_{k,l} - x_{k,l+1} \| = \| x_{k,l} - x_{k+1,0} \| \longrightarrow 0$$

Thus we may apply Theorem 6.3.1. Q.E.D.

COROLLARY 9.6.3. If the continuously differentiable, strongly quasi-convex functional f is *strictly pseudo-convex*—that is, if $\langle x - y, \nabla f(y) \rangle \geq 0$

implies $f(x) > f(y)$ for $x \neq y$—and if $\{x_{k,r}\}$ is bounded, then $x_{k,r} \longrightarrow \hat{x}$, the unique minimizer for f.

Proof: Limit points x' exist with $\nabla f(x') = 0$ by Corollary 9.6.1; then, by the strict pseudo-convexity, for any x, we have $0 = \langle x - x', \nabla f(x') \rangle$ and hence $f(x) > f(x')$, so x' minimizes f. By the strong quasi-convexity, such a minimizer is unique. Q.E.D.

COROLLARY 9.6.4. If f is uniformly quasi-convex, strictly pseudo-convex, and continuously differentiable, then for any $x_{1,0}$, the sequence $\{x_{k,r}\}$ converges to the unique $\hat{x}$ minimizing f.

COROLLARY 9.6.5 [Zangwill (1967)]. If $0 < aI \leq f''_x$ in $\mathbb{R}^l$, then for any $x_{1,0}$, the sequence $\{x_{k,r}\}$ converges to the unique $\hat{x}$ minimizing f.

EXERCISE. Prove Corollaries 9.6.4 and 9.6.5.

We have not, however, been able to prove that the method is exact for quadratics; Zangwill (1967) has developed a modification of the method that is exact. The *Zangwill method* is as follows.

Let e_i, $i = 1, \ldots, l$ be the coordinate directions and let an initial point $x_{0,l}$ and directions $d_{1,1}, \ldots, d_{1,l}$, $\|d_{1,i}\| = 1$ be given. Let $t_{0,l}$ minimize $f(x_{0,l} + td_{1,l})$ and let

$$x_{0,l+1} = x_{0,l} + t_{0,l}d_{1,l}$$

Set $n = 1$ and iteratively apply the *basic k-iteration* starting with $k = 1$; the basic k-iteration is as follows, given $x_{k-1,l+1}, d_{k,1}, \ldots, d_{k,l}$ and n: (1) compute α' to minimize $f(x_{k-1,l+1} + \alpha e_n)$, let $n' = n$, and replace n by $n(\text{modulo } l) + 1$. If $\alpha' \neq 0$, let $x_{k,0} = x_{k-1,l+1} + \alpha' e_{n'}$; if, however, $\alpha' = 0$, return to the start of step 1, noting that if step 1 is performed l times, we may consider $x_{k-1,l+1}$ to be the solution. (2) For $r = 1, \ldots, l$, compute $t_{k,r}$ to minimize $f(x_{k,r-1} + td_{k,r})$ and let

$$x_{k,r} = x_{k,r-1} + t_{k,r}d_{k,r}$$

Define

$$d_{k,l+1} = \frac{x_{k,l} - x_{k-1,l+1}}{\|x_{k,l} - x_{k-1,l+1}\|}$$

compute $t_{k,l+1}$ to minimize $f(x_{k,l} + td_{k,l+1})$, and set

$$x_{k,l+1} = x_{k,l} + t_{k,l+1}d_{k,l+1}$$

Define directions $d_{k+1,r} \equiv d_{k,r+1}$ for $r = 1, 2, \ldots, l$.

This method differs from the preceding primarily in its feature of minimizing over the coordinate directions as well as the directions $d_{k,r}$; this feature allows us to revise the directions $d_{k,r}$ in the simple manner of the algorithm and thus obtain exact convergence for quadratics, as the following theorem shows.

THEOREM 9.6.3 [Zangwill (1967)]. Let $f(x) = \langle h - x, M(h - x)\rangle$, where M is self-adjoint and positive-definite, and let the initial point $x_{0,l}$ be given. Then the iteration stops during step 1, with $x_{k-1,l+1} = h$, the solution, for some $k \leq l$.

Proof: Assume that at the start of the basic k-iteration at step k for $k \leq n - 1$, the directions

$$d_{k,l-k+1}, d_{k,l-k+2}, \ldots, d_{k,l}$$

are mutually M-conjugate and linearly independent; clearly this is true for $k = 1$, starting the induction. If we do not stop in step 1 this time, then $x_{k-1,l+1} \neq x_{k,0}$ and, since M is positive-definite, $f(x_{k,0}) < f(x_{k-1,l+1})$; but then also

$$f(x_{k,l}) \leq f(x_{k,0}) < f(x_{k-1,l+1})$$

so $x_{k-1,l+1} \neq x_{k,l}$ and hence $d_{k,l+1} \neq 0$. At iteration $k - 1$, since $d_{k-1,r+1} = d_{k,r}$, the last k directions to be used were $d_{k,l-k+1}, \ldots, d_{k,l}$. Since these directions were assumed to be linearly independent and M-conjugate, the point $x_{k-1,l+1}$ minimizes f over a k-dimensional flat spanned by those directions; similarly, however, $x_{k,l}$ yields the minimum over a parallel flat. Therefore, $d_{k,l+1}$ is M-conjugate to each of the directions $d_{k,l-k+1}, \ldots, d_{k,l}$. Since none of these directions nor $d_{k,l+1}$ is zero and M is positive-definite, it follows that $d_{k,l-k+1}, \ldots, d_{k,l}, d_{k,l+1}$ are linearly independent and mutually conjugate. Since these directions equal the directions

$$d_{k+1,l-(k+1)+1}, \ldots, d_{k+1,l}$$

the induction is complete. Thus, if the procedure has not stopped by the time we reach $k = l$, then the directions $d_{l,1}, \ldots, d_{l,l}$ are linearly independent and M-conjugate. In step 2 of the iteration with $k = l - 1$, we have minimized successively over these l directions and, therefore, $x_{l-1,l+1} = h$; therefore, we shall stop automatically during step 1 of the iteration with $k = l$. Q.E.D.

Thus this new method is exact for quadratics; it also is convergent for more general functionals, as was the preceding method. Since the *Proof* of the following is nearly the same as that for Theorem 9.6.2, we shall be very brief.

THEOREM 9.6.4. Let f be a continuous and strongly quasi-convex functional on $\mathbb{R}^l$, and let the new method of Zangwill described above be used, starting with an intitial point $x_{0,l}$ and directions $d_{1,1}, \ldots, d_{1,l}$; suppose that the sequence $\{x_{k,r}\}$ is bounded. Then any limit point x' of x_{k,r_0} for any fixed r_0 is also a limit point of $x_{k,r}$ for $r \neq r_0$, and for each such limit point x', we have $f(x') \leq f(x' + te_r)$ for all t and $r = 1, \ldots, l$, where the $\{e_r\}$ are the coordinate directions.

Proof: For any subsequence of integers K, we can take a further subsequence K_1 such that

$$\lim_{k \in K_1} x_{k+j,r} = x'_{j,r} \quad \text{for} \quad r = 0, 1, \ldots, l+1 \quad \text{and} \quad j = 0, 1, \ldots, l-1$$

Arguing precisely as in the *Proof* of Theorem 9.6.2, if necessary taking a further subsequence so that the directions also converge, we can finally conclude that there is a point x' with

$$x' = x'_{j,r} \quad \text{for} \quad r = 0, 1, \ldots, l+1 \quad \text{and} \quad j = 0, 1, \ldots, l-1$$

Since we have been considering the results of l consecutive basic k-iterations by dealing with the $x_{k+j,r}$ for $j = 0, 1, \ldots, l-1$, all l of the coordinate directions are used in the minimizations for each value of k as j and r vary. Arguing for these directions e_r as we did for the directions $d_1, \ldots, d_l$ in the *Proof* of Theorem 9.6.2, we conclude that $f(x') \leq f(x' + te_r)$ for all t and $r = 1, 2, \ldots, l$. Q.E.D.

Since the conclusions of this theorem for the new Zangwill method are effectively the same as for the old method in Theorem 9.6.2, all the corollaries of that theorem follow immediately. Thus we have the following corollary.

COROLLARY 9.6.6. Corollaries 9.6.1, 9.6.2, 9.6.3, 9.6.4, and 9.6.5 are valid for $\{x_{k,r}\}$ generated by the new Zangwill method above.

The two methods of Theorems 9.6.2 and 9.6.3, which attempt to generate *conjugate* directions without dealing with derivatives in any way whatsoever, appear to be about the best of the methods of this section, all of which ignore derivatives entirely.

EPILOGUE

We wish to close with some brief comments concerning the last four chapters of the text. The reader will have observed that we have not been able to give concrete statements as to which methods are the best to use; this is not only because there is no one "best" method, but also, and as importantly, because there have not been any really carefully performed tests, comparisons, and documented results for the many types of methods available for use on large classes of problems. Such a thorough study has been made [Engeli et al. (1959)] for the special case of convex quadratics, but nothing comparable exists for more general classes of problems. The few numerical examples we have given throughout the text have been taken from the literature and serve to indicate the difficulty of comparing results computed by different authors. Although no really thorough comparative studies of methods have been made, the reader can find a number of numerical examples for various methods in Broyden (1965), Fletcher (1965, 1968), Fletcher–Powell (1963), Fletcher–Reeves (1964), Goldstein–Price (1967), Kowalik–Osborne (1969), Powell (1964a, 1964b, 1968, 1969), and Stewart (1967). There is still a great need for readily available documentation and thorough comparison of those algorithms which appear to be useful, and until such data are available we can only give general guidelines based on incomplete data.

Some rough guidelines are as follows. It is generally best to use a conjugate-direction (especially conjugate-gradient), Newton-like, or even Newton method if this is feasible; since algorithms are available which exhibit the properties of these methods and which are designed to use various numbers of derivatives, some method of this type is almost always suitable for any given problem, and should be preferred. Generally, one need not try too hard to perform an exact minimization along a line. If the theoretical success of the method depends on the *exact* minimization, then some effort should be made; but in the other cases, function *reduction* is usually more important than function *minimization*, and some such method as in Section 4.5 or 4.6 might best be used. Admittedly, these are rather sparse guidelines, but until more valid comparisons are performed on significant classes of problems, little more can safely be said.

REFERENCES

ABADIE, J. (ed.) (1967), *Methods of nonlinear programming*, North Holland, Amsterdam.

AKAIKE, H. (1959), "On a successive transformation of probability distribution and its application to the analysis of the optimum gradient method," *Ann. Inst. Statist. Math.*, Tokyo, vol. 11, 1–16.

AKHIEZER, N. (1962), *The calculus of variations*, Blaisdell, Waltham, Mass.

ALTMAN, M. (1966a), "Generalized gradient methods of minimizing a functional," *Bull. Acad. Polon. Sci.*, vol. 14, 313–318.

ALTMAN, M. (1966b), "A generalized gradient method for the conditional minimum of a functional," *Bull. Acad. Polon. Sci.*, vol. 14, 445–451.

ANSELONE, P. M. (ed.) (1964), *Nonlinear integral equations*, Univ. of Wisc. Press, Madison.

ANSELONE, P. M. (1965), "Convergence and error bounds for approximate solutions of integral and operator equations," in *Error in digital computation*, vol. 2, ed. L. B. Rall, Wiley, N. Y.

ANSELONE, P. M. (1967), "Collectively compact operator approximations," Comput. Sci. Report #76, Stanford Univ., Stanford, Calif. Also to be published in Prentice-Hall series on Automatic Computation.

ANSELONE, P. M., R. H. MOORE (1964), "Approximate solutions of integral and operator equations," *J. Math. Anal. Appl.*, vol. 9, 268–277.

ANTOSIEWICZ, H. A., W. C. RHEINBOLDT (1962), "Conjugate direction methods and the method of steepest descent," 501–512, in *A survey of numerical analysis*, ed. J. Todd, McGraw-Hill, N. Y.

ARMIJO, L. (1966), "Minimization of functions having Lipschitz continuous first partial derivatives," *Pacific J. Math.*, vol. 16, 1–3.

AUBIN, J.-P. (1967a), "Approximation des espaces de distributions et des opérateurs différentiels," *Bull. Soc. Math. France Mémoire 12*, 1–139 (French).

AUBIN, J.-P. (1967b), "Behavior of the error of the approximate solutions of boundary value problems," *Ann. Scu. Norm. Pisa*, vol. 21, 599–637.

AUBIN, J.-P. (1968), "Evaluation des erreurs de troncations des approximations des espaces de Sobolev," *J. Math. Anal. Appl.*, vol. 21, 356–368 (French).

AUBIN, J.-P. (1969a), "Estimate of the error in the approximation of optimization problems with constraints by problems without constraints," 153–175, in *Control theory and the calculus of variations*, ed. A. V. Balakrishnan, Academic, N. Y.

AUBIN, J.-P. (1969b), private communication.

AUBIN, J.-P., J. L. Lions (1966), unpublished notes on minimization, private communication.

BALAKRISHNAN, A. V., L. W. NEUSTADT (1964), *Computing methods in optimization problems*, Academic, N. Y.

BALUEV, A. (1952), "On the method of Chaplygin," *Dokl. Akad. Nauk CCCP*, vol. 83, 781–784 (Russian).

BARNES, J. (1965), "An algorithm for solving nonlinear equations based on the secant method," *Comput. J.*, vol. 8, 66–72.

BELLMAN, R. (1957), "On monotone convergence to solutions of $u' = g(u, t)$," *Proc. AMS*, vol. 8, 1007–1009.

BELLMAN, R. (1962), "Quasi-linearization and upper and lower bounds for variational problems," *Quart. Appl. Math.*, vol. 19, 349–350.

BEN-ISRAEL, A. (1965), "A modified Newton–Raphson method for the solution of equations," *Israel J. Math.*, vol. 3, 94–98.

BEN-ISRAEL, A. (1966), "A Newton-Raphson method for the solution of systems of equations," *J. Math. Anal. Appl.*, vol. 15, 243–252.

BERS, L. (1953), "On mildly nonlinear partial difference equations of elliptic type," *J. Res. Nat. Bur. Standards*, vol. 51, 229–236.

BROWDER, F. E. (1967), "Approximation-solvability of nonlinear functional equations in normed linear spaces," *Arch. Rat. Mech. Anal.*, vol. 26, 33–42.

BROYDEN, C. G. (1965), "A class of methods for solving nonlinear simultaneous equations," *Math. Comp.*, vol. 19, 577–593.

BROYDEN, C. G. (1967), "Quasi-Newton methods and their application to function minimization," *Math. Comp.*, vol. 21, 368–381.

BUDAK, B. M., E. M. BERKOVICH, E. N. SOLOVEVA, (1968–69), "Difference approximations in optimal control problems," *Vest. Mosc. Univ. Ser. Mat. Mech.* (1968), 41–55 (Russian), and translated in *SIAM J. Control*, vol. 7 (1969), 18–31.

BUTLER, T., A. MARTIN (1962), "On a method of Courant for minimizing functionals," *J. Math. Phys.*, vol. 41, 291–299.

CASPAR, J. R. (1968), "Applications of alternating direction methods to mildly nonlinear problems," Comput. Sci. Report # 68–83, Univ. of Md., College Park.

CAUCHY, A. (1847), "Méthodes générales pour la résolution des systèmes d'équations simultanées," *C. R. Acad. Sci. Par.*, vol. 25, 536–538 (French).

CEA, J. (1969), Lectures on minimization problems, l'Ecole d'Eté Analyse Numérique, France (French).

CIARLET, P. G. (1966), "Variational methods for nonlinear boundary value problems," dissertation, Case Inst. Tech., Cleveland.

CIARLET, P. G., M. H. Schultz, R. S. Varga (1967), "Numerical methods of high order accuracy for nonlinear boundary value problems I: One dimensional problems," *Numer. Math.*, vol. 9, 394–430.

CIARLET, P. G., M. H. SCHULTZ, R. S. VARGA (1968a), "Numerical methods of high order accuracy for nonlinear boundary value problems II: Nonlinear boundary conditions," *Numer. Math.*, vol. 11, 331–345.

CIARLET, P. G., M. H. SCHULTZ, R. S. VARGA (1968b), "Numerical methods of high order accuracy for nonlinear boundary value problems IV: Periodic boundary conditions," *Numer. Math.*, vol. 12, 266–279.

COLLATZ, L. (1966), *Functional analysis and numerical mathematics*, Academic, N. Y.

COURANT, R. (1943), "Variational methods for the solution of problems of equilibrium and vibrations," *Bull. AMS*, vol. 49, 1–23.

COURANT, R., D. HILBERT (1953), *Methods of mathematical physics*, vol. 1, Interscience, N. Y.

CROSS, K. E. (1968), "A gradient projection method for constrained optimization," Union Carbide Nuclear Div. Report # K-1746.

CULLUM, J. (1969), "Discrete approximations to continuous optimal control problems," *SIAM J. Control*, vol. 7, 32–50.

CULLUM, J. (1970), "An explicit procedure for discretizing continuous optimal control problems," to appear in *J. Opt. Th. Appl.*

CURRY, H. B. (1944), "The method of steepest descent for nonlinear minimization problems," *Quart. Appl. Math.*, vol. 2, 258–263.

DANIEL, J. W. (1965), "The conjugate gradient method for linear and nonlinear operator equations," dissertation, Stanford Univ., Stanford, Calif.

DANIEL, J. W. (1967a), "Convergence of the conjugate gradient method with computationally convenient modifications," *Numer. Math.*, vol. 10, 125–131.

DANIEL, J. W. (1967b), "The conjugate gradient method for linear and nonlinear operator equations," *SIAM J. Numer. Anal.*, vol. 4, 10–26.

DANIEL, J. W. (1968a), "Collectively compact sets of gradient mappings," *Indag. Math.*, vol. 30, 270–279.

DANIEL, J. W. (1968b), "On the approximate minimization of functionals," Comput. Sci. Report # 42, Univ. of Wisc., Madison, also in *Math. Comp.*, vol. 23, 573–582.

DANIEL, J. W. (1969), "Convergence of the conjugate gradient method: a correction," to appear in *SIAM J. Numer. Anal.*

DANIEL, J. W. (1970), "Convergence of a discretization for constrained spline function problems," Comput. Sci. Report # 76, Univ. of Wisc., Madison.

DAVIDON, W. C. (1959), "Variable metric method for minimization," A. E. C. Res. and Dev. Report # ANL-5990.

DAVIDON, W. C. (1968), "Variance algorithm for minimization," *Comput. J.*, vol. 10, 406–410.

DEMYANOV, V. F., A. M. Rubinov (1967), "The minimization of a smooth convex functional on a convex set," *J. SIAM Control*, vol. 5, 280–294.

DENNIS, J. E., Jr. (1969), "On the convergence of Newton-like methods," presented at Conf. on Numer. Meth. Nonlinear Algebraic Equations, Univ. of Essex.

DI GUGLIELMO, F. (1969), "Construction d'approximations des espaces de Sobolev $H^m(\mathbb{R}^n)$ sur des résaux en simplexes," to appear (French).

DOUGLAS, J. E., Jr., R. B. KELLOGG, R. S. VARGA (1963), "Alternating direction methods for n space variables," *Math. Comp.*, vol. 17, 279–282.

DUNFORD, N., J. Schwartz (1962), *Linear operators I: General theory*, Interscience, N. Y.

ELKIN, R. M. (1968), "Convergence theorems for Gauss–Seidel and other minimization algorithms," Comput. Sci. Report # 68–59, Univ. of Md., College Park.

ENGELI, M., TH. GINSBURG, H. RUTISHAUSER, E. STIEFEL (1959), "Refined iterative methods for computation of the solution and the eigenvalues of self-adjoint boundary value problems," Mitt. Inst. Angew. Math., Zurich, # 8.

FADDEEV, D. K., V. N. FADDEEVA (1963), *Computational methods of linear algebra*, Freeman, San Francisco.

FIACCO, A., G. MCCORMICK (1968), *Nonlinear programming: sequential unconstrained minimization techniques*, Wiley, N. Y.

FIX, G., G. STRANG (1969), "Fourier analysis of the finite element method in Ritz–Galerkin theory," to appear.

FLETCHER, R. (1965), "Function minimization without evaluating derivatives, a review," *Comput. J.*, vol. 8, 33–41.

FLETCHER, R. (1968), "Generalized inverse methods for the best least squares solution of systems of nonlinear equations," *Comput. J.*, vol. 10, 392–399.

FLETCHER, R., M. Powell (1963), "A rapidly convergent descent method for minimization," *Comput. J.*, vol. 6, 163–168.

FLETCHER, R., C. Reeves (1964), "Function minimization by conjugate gradients," *Comput. J.*, vol. 7, 149–154.

FORSYTHE, G. E. (1968), "On the asymptotic directions of the s-dimensional optimum gradient method," *Numer. Math.*, vol. 11, 57–76.

FRANK, M., P. WOLFE (1956), "An algorithm for quadratic programming," *Nav. Res. Log. Quar.*, vol. 3, 95–110.

GILBERT, E. G. (1966), "An iterative procedure for computing the minimum of a quadratic form on a convex set," *SIAM J. Control*, vol. 4, 61–81.

GOLDFARB, D. (1966), "A conjugate gradient method for nonlinear programming," dissertation, Princeton Univ., Princeton, N. J.

GOLDFARB, D. (1969a), "Extension of Davidon's variable metric method to maximization under linear inequality and equality constraints," *SIAM J. Appl. Math.*, vol. 17, 739–764.

GOLDFARB, D. (1969b), "Sufficient conditions for the convergence of a variable metric algorithm," 273–282, in *Optimization*, ed. R. Fletcher, Academic, N. Y.

GOLDFARB, D., L. LAPIDUS (1968), "Conjugate gradient method for nonlinear programming problems with linear constraints," *I. and E. C. Fundamentals*, vol. 7, 142–151.

GOLDSTEIN, A. A. (1964a), "Convex programming in Hilbert space," *Bull. AMS*, vol. 70, 709–710.

GOLDSTEIN, A. A. (1964b), "Minimizing functionals on Hilbert space," 159–166, in Balakrishnan and Neustadt (1964).

GOLDSTEIN, A. A. (1965), "On steepest descent," *J. SIAM Control*, vol. 3, 147–151.

GOLDSTEIN, A. A. (1966), "Minimizing functionals on normed linear spaces," *J. SIAM Control*, vol. 4, 81–89.

GOLDSTEIN, A. A. (1967), *Constructive real analysis*, Harper & Row, N. Y.

GOLDSTEIN, A. A., J. F. PRICE (1967), "An effective algorithm for minimization," *Numer. Math.*, vol. 10, 184–189.

GOLUB, G. H., M. SAUNDERS (1969), "Linear least squares and quadratic programming," Comput. Sci. Report CS 134, Stanford Univ., Stanford, Calif.

GREENSPAN, D. (1965), "On approximating extremals of functionals I," *ICC Bull.*, vol. 4, 99-120.

GREENSPAN, D. (1967), "On approximating extremals of functionals II," *Int. J. Eng. Sci.*, vol. 5, 571–588.

GREENSPAN, D., S. PARTER (1965), "Mildly nonlinear elliptic partial differential equations and their numerical solution II," *Numer. Math.*, vol. 7, 129–145.

HADLEY, G. (1964), *Nonlinear and dynamic programming*, Addison-Wesley, Reading, Mass.

HAYES, R. M. (1954), "Iterative methods of solving linear problems in Hilbert space," 71–104, in *Contributions to the solution of systems of linear equations and the determination of eigenvalues*, ed. O. Taussky, vol. 39, Nat. Bur. Standards Appl. Math. Ser.

HENRICI, P. (1962), *Discrete variable methods in ordinary differential equations*, Wiley, N. Y.

HERBOLD, R. J. (1968), "Consistent quadrature schemes for the numerical solution of boundary value problems by variational techniques," dissertation, Case Western Reserve, Cleveland.

HERBOLD, R., M. SCHULTZ, R. VARGA (1969), "Quadrature schemes for the numerical solution of boundary value problems by variational techniques," Comput.

Sci. Report, Carnegie-Mellon Univ., Pittsburgh, Pa. Also in *Aeq. Math.*, vol. 3, 96–119.

HESTENES, M. R. (1956), "The conjugate gradient method for solving linear systems," *Proc. Symp. Appl. Math.*, vol. 6 (*Num. Anal.*), 83–102.

HESTENES, M. R. (1966), *Calculus of variations and optimal control theory*, Wiley, N. Y.

HESTENES, M. R., E. STIEFEL (1952), "Method of conjugate gradients for solving linear systems," *J. Res. Nat. Bur. Standards*, vol. 49, 409–436.

HOOKE, R., T. JEEVES (1961), "Direct search solution of numerical and statistical problems," *JACM*, vol. 8, 212–229.

HORWITZ, L. B., P. E. SARACHIK (1968), "Davidon's method in Hilbert space," *SIAM J. Appl. Math.*, vol. 16, 676–696.

HUARD, P. (1967), "Resolution of mathematical programming with nonlinear constraints by the method of centres," 209–219, in Abadie (1967).

KANTOROVICH, L. V. (1948), "Functional analysis and applied mathematics," *Uspekhi Mat. Nauk*, vol. 3, 89–185 (Russian). Also translated as Nat. Bur. of Standards Report # 1509 (1952).

KANTOROVICH, L. V., G. P. AKILOV (1964), *Functional analysis in normed linear spaces*, Macmillan, N. Y.

KANTOROVICH, L. V., V. I. KRYLOV (1958), *Approximate methods of higher analysis*, Noordhoff, Groningen, Netherlands.

KAUPE, A. F., Jr. (1963), Algorithm 178, "Direct search," *CACM*, vol, 6, 313, and *CACM*, vol. 9, 684.

KELLOGG, R. (1969), "A nonlinear alternating direction method," *Math. Comp.*, vol. 23, 23–27.

KIEFER, J. (1957), "Optimum sequential search and approximation methods under minimum regularity assumptions," *SIAM J. Appl. Math.*, vol. 5, 105–136.

KOWALIK, J., M. OSBORNE (1968), *Methods for unconstrained optimization problems*, American Elsevier, N. Y.

KRABS, W. (1963), "Einige Methoden zur Losung des diskreten linearen Tscheby-scheff-Problems," dissertation, Univ. of Hamburg (German).

KRASOVSKII, N. N. (1957), "On a problem in optimal control," *Prik. Mat. Mex.*, vol. 21, 670–677 (Russian).

KREUSER, J. (1969), private communication.

LANGENHOP, C. E. (1967), "On generalized inverses of matrices," *SIAM J. Appl. Math.*, vol. 15, 1239–1246.

LEVENBERG, K. (1944), "A method for the solution of certain nonlinear problems in least squares," *Quart. Appl. Math.*, vol. 2, 164–168.

LEVITIN, E. S., B. T. POLJAK (1966a), "Constrained minimization methods," *Zh. vych. Mat. mat. Fiz.*, vol. 6, 787–823 (Russian). Also translated in *USSR Comput. Math. Math. Phys.*, vol. 6 (1968), 1–50.

LEVITIN, E. S., B. T. POLJAK (1966b), "Convergence of minimizing sequences in conditional extremum problems," *Soviet Math. Dokl.*, vol. 7, 764–767.

LIEBERSTEIN, H. (1959), "Over-relaxation for nonlinear elliptic partial differential equations," Math. Res. Cntr. Report # 80, Madison, Wisc.

LORENTZ, G. G. (1966), *Approximation of functions*, Holt, N. Y.

MANGASARIAN, O. L. (1969), *Nonlinear programming*, McGraw-Hill, N. Y.

MARQUARDT, D. W. (1963), "An algorithm for least squares estimation of nonlinear parameters," *SIAM J. Appl. Math.*, vol. 11, 431–441.

MCCORMICK, G. P. (1969), private communication.

MIKHLIN, S. G., K. L. SMOLITSKIY (1967), *Approximate methods for solution of differential and integral equations*, American Elsevier, N. Y.

MOORE, R. H. (1966), "Differentiability and convergence for compact nonlinear operators," *J. Math. Anal. Appl.*, vol. 16, 65-72.

MORREY, C. B., JR. (1966), *Multiple integrals in the calculus of variations*, Springer-Verlag, N. Y.

MORRISON, D. D. (1960), "Methods for nonlinear least squares problems and convergence proofs," JPL Seminar Proc., Space Tech. Lab., Los Angeles.

MYERS, G. E. (1968), "Properties of the conjugate gradient and Davidon methods," *J. Opt. Th. Appl.*, vol. 2, 209–219.

NASHED, M. Z. (1967), "Supportably and weakly convex functionals with applications to approximation theory and nonlinear programming," *J. Math. Anal. Appl.* vol. 18, 504–521.

NOBLE, B. (1964), "Complementary variational principles for boundary value problems I: Basic principles," Math. Res. Cntr. Report # 473, Madison, Wisc.

NOBLE, B. (1966), "Complementary variational principles for boundary value problems II: Nonlinear networks," Math. Res. Cntr. Report # 643, Madison, Wisc.

ODLOLESKAL, L. (1969), "Questions of accuracy and effectiveness of computing methods," Trudy Symp., Kiev, vol. 4, 94-103 (Russian).

ORTEGA, J., W. C. RHEINBOLDT (1967a), "Monotone iterations for nonlinear equations with applications to Gauss–Seidel methods," *SIAM J. Numer. Anal.*, vol. 4, 171–190.

ORTEGA, J., W. C. RHEINBOLDT (1967b), "On a class of approximate iterative processes," *Arch. Rat. Mech. Anal.*, vol. 3, 352–365.

ORTEGA, J. M., W. C. RHEINBOLDT (1968), "Local and global convergence of generalized linear iterations," Comput. Sci. Report # 68–82, Univ. of Md., College Park.

ORTEGA, J., M. ROCKOFF (1966), "Nonlinear difference equations and Gauss–Seidel type iterative methods," *SIAM J. Numer. Anal.*, vol. 3, 497–513.

OSTROWSKI, A. M. (1966a), "Contributions to the theory of the method of steepest descent I," Math. Res. Cntr. Report # 615, Madison, Wisc.

OSTROWSKI, A. M. (1966b), *Solution of equations and systems of equations*, Academic, N. Y.

PENROSE, R. (1955), "A generalized inverse for matrices," *Proc. Camb. Phil. Soc.*, vol. 51, 406–413.

PEREYRA, V. (1967), "Iterative methods for solving nonlinear least squares problem," *SIAM J. Numer. Anal.*, vol. 4, 27–36.

PERR′N, F. M., H. S. PRICE, R. S. VARGA (1969), "On higher order numerical methods for nonlinear two-point boundary value problems," *Numer. Math.*, vol. 13, 180–198.

PETRYSHYN, W. V. (1968), "On the approximation-solvability of nonlinear equations," *Math. Ann.*, vol. 177, 156–164.

POLJAK, B. T. (1966), "Existence theorems and convergence of minimizing sequences in extremum problems with restrictions," *Soviet Math. Dokl.*, vol. 7, 72–75.

POLJAK, B. T. (1969a), "Conjugate gradient method in extremal problems," Works of the All-Union School on Mathematical Programming (Russian), also in *J. Comput. Math. and Math. Phys.*, vol. 9, 809–821 (Russian).

POLJAK, B. T. (1969b), "Iterative methods of solving some incorrect variational problems," *Computing methods and programming*, vol. 12, Moscow State Univ., 38–52 (Russian).

POLJAK, B. T. (1969c), "On one method of solving large size linear and quadratic programming problems," *Computing methods and programming*, vol. 12, Moscow State Univ., 10–17 (Russian).

POLJAK, B. T., T. P. Ivanov, G. V. PUKOV (1967), "Numerical methods of solving some extremal problems with partial derivatives," *Computing methods and programming*, vol. 9, Moscow State Univ., 194–203 (Russian).

POLJAK, B. T., V. S. ORLOV, V. A. REBRY, N. V. TRETJAKOV (1967), "Experimental solution of the problem of optimal control," *Computing methods and programming*, vol. 9, Moscow State Univ., 179–193 (Russian).

POLJAK, B. T., V. A. SKOKOV (1967a), "Choice of parameters of kinetic equations with respect to experimental data," *Computing methods and programming*, vol. 9, Moscow State Univ., 167–178 (Russian).

POLJAK, B. T., V. A. SKOKOV (1967b), "Standard programs for minimizing functions of several variables (for the machine M-20)," in "Standard programs for solving mathematical programming problems," vol. 4, Moscow State Univ. (Russian).

PONSTEIN, J. (1967), "Seven types of convexity," *SIAM Rev.*, vol. 9, 115–119.

PONTRYAGIN, L. S., V. G. BOLTYANSKII, R. V. GAMKRILIDZE, E. F. MISCENKO (1962), *The mathematical theory of optimal processes*, Wiley, N. Y.

POWELL, M. J. (1964a), "A method for minimizing a sum of squares of nonlinear functions without calculating derivatives," *Comput. J.*, vol. 7, 303–307.

POWELL. M. J. (1964b), "An efficient method for finding the minimum of a function of several variables without calculating derivatives," *Comput. J.*, vol. 7, 155–162.

POWELL, M. J. (1968), "A FORTRAN subroutine for solving systems of nonlinear algebraic equations," Unit. King. Atom. Energy Auth. Res. Grp. Report # R5947, Harwell, Berkshire.

POWELL, M. J. (1969), "A hybrid method for nonlinear equations," to appear.

POWELL, M. J. (1970), "On the convergence of the variable metric algorithm," to appear.

PROTTER, M. H., H. F. WEINBERGER (1967), *Maximum principles in differential equations*, Prentice-Hall, Englewood Cliffs, N. J.

RABINOWITZ, P. (1968), "Applications of linear programming to numerical analysis," *SIAM Rev.*, vol. 10, 121–160.

RALL, L. (1966), "On complementary variational principles," *J. Math. Anal. Appl.*, vol. 14, 174–184.

RALL, L. B. (1969,) *Computational solution of nonlinear operator equations*, Wiley, N. Y.

RHEINBOLDT, W. C. (1967), "On a unified convergence theory for a class of iterative processes," Comput. Sci. Report #TR-67-46, Univ. of Md., College Park.

RITTER, K. (1969), private communication.

RIVLIN, T., W. CHENEY (1966), "A comparison of uniform approximation on an interval and a finite subset thereof," *SIAM J. Numer. Anal.*, vol. 3, 311–320.

ROSEN, J. B. (1960–61), "The gradient projection method for nonlinear programming I: Linear constraints," *J. SIAM*, vol. 8, 181–217; "The gradient projection method for nonlinear programming II: Nonlinear constraints," *J. SIAM*, vol. 9, 514–532.

ROSEN, J. B. (1966), "Iterative solution of nonlinear optimal control problems," *J. SIAM Control.* vol. 4, 223–244.

ROSEN, J. B. (1968), "Approximate solution and error bounds for quasi-linear elliptic boundary value problem," Comput. Sci. Report # 30, Univ. of Wisc., Madison.

ROSEN, J., R. MEYER (1967), "Solution of nonlinear two point boundary value problems by linear programming," Comput. Sci. Report #1, Univ. of Wisc., Madison.

ROSENBROCK, G. H. (1960), "An automatic method for finding the greatest or the least value of a function," *Comput. J.*, vol. 3, 175–184.

ROXIN, E. (1962), "The existence of optimal controls." *Mich. Math. J.*, vol. 9, 109–119.

SCHECHTER, S. (1962), "Iteration methods for nonlinear problems," *Trans. AMS*, vol. 104, 179–189.

SCHECHTER, S. (1968), "Relaxation methods for convex problems," *SIAM J. Numer. Anal.*, vol. 5, 601–612.

SHAMPINE, L. (1966), "Monotone iterations and two-sided convergence," *SIAM J. Numer. Anal.*, vol. 3, 607–615.

SHAMPINE, L. (1968), "Error bounds and variational methods for nonlinear boundary value problem," *Numer. Math.*, vol. 12, 410–415.

SIMPSON, R. B. (1968), "Approximation of the minimizing element for a class of functionals," *SIAM J. Numer. Anal.*, vol. 5, 26–41.

SIMPSON, R. B. (1969) "The Rayleigh–Ritz process for the simplest problem in the calculus of variation," *SIAM J. Numer. Anal.*, vol. 5, 258–271.

SMITH, C. S. (1962), "The automatic computation of maximum likelihood estimates," N. C. B. Sci. Dept. Report #SC846/MR/40.

SOUTHWELL, R. V. (1946), *Relaxation methods in theoretical physics*, vol. I, Clarendon, Oxford.

SPANG, H. A., III (1962) "A review of minimization techniques for nonlinear functions." *SIAM Rev.*, vol. 4, 343–365.

STEWART, G. W., III (1967), "A modification of Davidon's minimization method to accept difference approximations of derivatives," *JACM*, vol. 14, 72–83.

STIEFEL, E. (1954), "Recent developments in relaxation techniques," *Proc. Int. Cong. Appl. Math.*, vol. I, 384–391.

STIEFEL, E. (1955), "Relaxationsmethoden bester Strategie zur Losung linearer Gleichungs-systeme," *Comment. Math Helv.*, vol. 29, 157–179.

SWANN, W. H. (1964), "Report on the development of a new direct search method of optimization," I. C. I., Ltd., Cent Instr. Lab. Res. Note #64/3, Wilmslow, England.

TAYLOR, A. E. (1961), *Introduction to functional analysis*, Wiley, N. Y.

TIKHONOV, A. N. (1965), "Methods for the regularization of optimal control problems," *Soviet Math. Dokl.*, vol. 6, 761—763.

TOPKIS, D. M., A. F. VEINOTT, Jr. (1967), "On the convergence of some feasible direction algorithms for nonlinear programming," *SIAM J. Control*, vol. 5, 268–279.

TRAUB, J. F. (1964), *Iterative methods for the solution of equations*, Prentice-Hall, Englewood Cliffs, N. J.

TREMOLIERES, R. (1969), private communication.

VAINBERG, M. M. (1964), *Variational methods for the study of nonlinear operators*, Holden-Day, San Francisco.

VARGA, R. S. (1962), *Matrix iterative analysis*, Prentice-Hall, Englewood Cliffs, N. J.

VERCOUSTRE, A.-M. (1969), "Critère général de convergence pour les méthodes de quasi-Newton," to appear.

WARGA, J. (1962), "Necessary conditions for a minimum in a relaxed variational problem," *J. Math. Anal. Appl.*, vol. 4, 129–145.

WASSCHER, E. J. (1963), Algorithms 203–204, "Steep 1" and "Steep 2," *CACM*, vol. 6, 517–519, *CACM*, vol. 7, 585, and *CACM*, vol. 8, 171.

WELLS, M. (1965), Algorithm 251, "Flepomin," *CACM*, vol. 8, 169 and *CACM*, vol. 9, 686.

WHITLEY, V. W. (1932), Algorithm 129, "Minifun," *CACM*, vol. 5, 550, and *CACM*, vol. 6, 521.

YAKOVLEV, M. N. (1965), "On some methods of solving nonlinear equations," *Trudy Mat. Inst. Steklov*, vol. 84, 8–40 (Russian). Translated as Comput. Sci. Report #68–75, Univ. of Md., College Park.

ZANGWILL, W. I. (1967), "Minimizing a function without calculating derivatives," *Comput. J.*, vol. 10, 293–296.

ZANGWILL, W. I. (1969), *Nonlinear programming: a unified approach*, Prentice-Hall, Englewood Cliffs, N. J.

ZELEZNIK, F. J. (1968), "Quasi-Newton methods for nonlinear equations," *JACM*, vol. 15, 265–271.

ZOUTENDIJK, G. (1960), *Methods of feasible directions*, Elsevier, Amsterdam.

INDEX

L

M

N

O

P

Q

R

S

T

U

V

W

Z